Probability & Evidence

COLUMBIA CLASSICS IN PHILOSOPHY

Probability & Evidence

A. J. Ayer

Introduction by Graham Macdonald

COLUMBIA UNIVERSITY PRESS • NEW YORK

COLUMBIA UNIVERSITY PRESS

Publishers Since 1893
New York   Chichester, West Sussex

Copyright © 1972 A. J. Ayer
Introduction © 2006 Columbia University Press
All rights reserved

Library of Congress Cataloging-in-Publication Data

Ayer, A. J. (Alfred Jules), 1910–
    Probability and evidence / A. J. Ayer ; introduction by Graham
Macdonald.
        p.   cm. — (Columbia classics in philosophy)
        Originally published: 1972, in series: John Dewey essays in
philosophy. With new introd. and bibliography.
        Includes bibliographical references and index.
        ISBN 0–231–13274–3 — ISBN 0–231–13275–1 (pbk.)
        1. Induction (Logic)   2. Probabilities.   3. Hume, David,
1711–1776.   4. Harrod, Roy Forbes, Sir, 1900– Foundations of
inductive logic.   I. Title.   II. Series.

    B1618.A93P76 2006
    121'.6—dc22

                                                        2005049776

♾ Columbia University Press books are printed on permanent
and durable acid-free paper.

Printed in the United States of America

c 10 9 8 7 6 5 4 3 2 1
p 10 9 8 7 6 5 4 3 2 1

To Margie & Goronwy Rees

Contents

Introduction

BIOGRAPHICAL SKETCH

Alfred Jules Ayer was born in London on October 29, 1910. His mother, Reine, was descended from Dutch Jews, and his father, Jules Louis Cypress Ayer, came from a Swiss Calvinist background. He began reading philosophical works fairly early in his life; by the age of sixteen he was reading Bertrand Russell's *Sceptical Essays,* which made a deep impression on him. He was particularly taken with Russell's claim that it is undesirable to believe a proposition when there is no ground for believing its truth. Ayer said that this remained a motto for him throughout his philosophical career.[1] It can be seen as the starting point for his lifelong quest to establish foundations for knowledge; this led him to do extensive work both on perceptual knowledge and on the legitimacy of inductive inferences.

In 1929 he won a classics scholarship to Christ Church, Oxford, where he studied Greek and philosophy. One of his tutors was Gilbert Ryle. Ryle was also instrumental in getting Ayer to go to Vienna in 1933 to study with Moritz Schlick, then leader of the Vienna Circle; Ayer joined W. V.O. Quine as one of only two foreign visitors to be members of the circle.

After leaving Vienna, Ayer lectured for a short time at Christ Church. In 1935, at the young age of twenty-five, he finished his first and most famous book, *Logic, Truth, and Language.* In it he defended the positivism he had imbibed while in Vienna, basing the rejection of all metaphysical theses on the principle of verification, which functioned as a criterion of the meaningfulness of statements. This book became widely known partly because of the controversy it caused: not only did Ayer claim that religious assertions, those supposedly enunciating truths about God, were meaningless, he also claimed the same about everyday moral

---

[1] Much of this biographical material is drawn from Ben Rogers, *A. J. Ayer* (London: Abacus Press, 1992).

assertions. For this he was lampooned in *Time* magazine as being a man
who would judge that believing one should help one's mother was
meaningless. In the years immediately after the publication of *Language,
Truth, and Logic* Ayer worked at defending and refining some of the positions adopted in that book, not least at meetings in Oxford with Isaiah
Berlin, Stuart Hampshire, and J. L. Austin. The product of this refining
process was the book *The Foundations of Empirical Knowledge*, published
in 1940.

Ayer's first 'official' encounter with America was during the war,
when he was sent there to join a secret service mission, which seems to
have involved gathering information about Fascist sympathizers. While
in New York he reviewed films for the *Nation*, fathered a daughter (by
Sheila Graham), befriended e. e. cummings and his wife, Marianne, and
made a record with Lauren Bacall. After the war Ayer tutored at Wad-
ham College, Oxford, but at thirty-six he moved from Oxford to become the Grote Professor of Philosophy at University College, London.
In 1948 he made the first of his many academic trips to the United
States, lecturing at Bard, then a college of Columbia University. In 1958
Ayer returned to Oxford as Wykeham Professor of Logic, where he
stayed until his retirement in 1978.

During the late 1960s and early 1970s Ayer was philosophically very
productive, doing some of his most original work. *The Origins of Pragmatism* was published in 1968, followed by *Russell and Moore: The Analytical Heritage* (1971, the product of the William James Lectures he delivered at Harvard in 1970), and *Probability and Evidence* (1972, the Dewey
Lectures delivered at Columbia University in 1970). Shortly thereafter
came *Russell*, a small paperback, and *The Central Questions of Philosophy*
(1973, originally given as the Gifford Lectures at the University of St.
Andrews), in which he elaborated on the sophisticated realism first put
forward in *The Origins of Pragmatism*.

After his retirement Ayer continued to enjoy philosophising and travelling. Some of his philosophical work involved replying to articles in
volumes published in his honour (Macdonald [1979], Macdonald and
Wright [1986], Honderich [1991], Hahn [1992]).[2] There was also a full-
length critical study by John Foster in the prestigious Routledge Arguments of the Philosophers series (Foster 1985). Ayer must have just
completed his replies to the Hahn collection before he was admitted

---

[2] Ayer's replies are to be found in Macdonald (1979) and Hahn (1992).

to hospital with a collapsed lung in early summer 1989, and he died on
June 27.

Ayer wrote two autobiographies, *Part of My Life* and *More of My
Life*. Both were packed with detail, but not with a great deal of passion
or self-scrutiny. His private life was eventful; he married four times
(twice to Dee Wells) and had numerous affairs. (Ben Rogers has written a sympathetic and insightful biography [Rogers 2002].) His circle
of friends included many famous and influential people; the following
(in no particular order) is only a brief list: Cyril Connolly, Graham
Greene, George Orwell, e. e. cummings and his wife, Marianne, Meyer
Schapiro, Arthur Koestler, Bertrand Russell, Stephen Spender, Wynston
Auden, Philip Toynbee, Isaiah Berlin, Hugh Gaitskell, Roy Jenkins, Mi-
chael Foot, Richard Crossman, Jonathon Miller, Angus Wilson, Alan
Bennett, Alice Thomas Ellis, Jane Fontaine, Iris Murdoch, V. S. Pritchett,
and Christopher Hitchens. He believed, maybe truly, that the character
of George Moore in Tom Stoppard's play *Jumpers* was modeled on him.
Among the honours bestowed on Ayer were his knighthood, Fellow
of the British Academy, honorary member of the American Academy
of Arts and Sciences, Member of the Bulgarian Order of Cyril and
Methodius, 1st class, and Chevalier of the Legion d'Honneur.

PROBABILITY AND EVIDENCE

Ben Rogers, Ayer's biographer, talks of a particularly important period
of Ayer's philosophical career as the 'constructionist phase'. It started
with *The Origins of Pragmatism* (a book dealing with Pierce and James,
but not Dewey), continued with *Russell and Moore* and *The Central Questions of Philosophy,* and finished with *Probability and Evidence*. Rogers
claims that *Probability and Evidence* is "the least constructive and for that
reason the least interesting of this cluster" (300), a judgment with which
I disagree. Problems relating to evidence and confirmation were right at
the heart of Ayer's philosophical concerns, starting with his attempts to
formulate the verification principle, a strict criterion of meaningfulness.
Essentially, the principle declared that for a statement to be meaningful
it must be directly or indirectly verifiable, and for this to be true it had
to be independently related to observation statements. Ayer struggled to
render the notion of 'indirect verifiability' watertight but didn't succeed

(see Macdonald [2005] for a more detailed account of the verification principle and a guide to some of the subsequent literature). He remained, though, an empiricist about meaning, claiming that meaningful statements have to 'make a difference' in what we can expect to observe. He said that the exact formulation of the verification principle would only be possible once we had a sound theory of confirmation and evidence.[3] *Probability and Evidence* is just such a sustained attempt to come to grips with the evidential relations between propositions.

This was not the only issue that exercised Ayer; against the conventionalism of Carnap and Popper, he held the iconoclastic view that our *experience* gives us reason to believe certain observation statements (conventionalism being the thesis that these observations are accepted 'by agreement', and are not justified by experience). In his early work he held that our beliefs about the existence of physical objects were justified directly by our sense-data, because the physical objects were just collections of (actual and possible) sense-data. Ayer defended this phenomenalism in *Language, Truth, and Logic* and *The Foundations of Empirical Knowledge*.

The work done in *The Origins of Pragmatism* and *Russell and Moore* was primarily historical, with the original section in *The Origins* having to do with Ayer's newly elaborated theory of perception, which he gave a fuller hearing in *The Central Questions of Philosophy*. This theory also dealt with how our knowledge of the physical world was constructed from our perceptual experience, still seen as the foundation for all our knowledge. What was new in these books was the rejection of phenomenalism in favour of a nonreductive realism about physical objects; they were no longer seen as mere collocations of actual and possible sense-data. On this new theory (Ayer called it 'sophisticated realism' [Ayer 1973]), our perception of physical objects was held to be indirect, given that the ultimate basis for perceptual judgments remained sense-data, now called 'quale'. Ayer now held that the patterns in our experience formed by quale constituted a primary system, on the basis of which we posited the existence of physical objects, the system of physical objects being the 'theoretical' secondary system. It was only when the secondary system was posited that we were able to make distinctions between what was mental and what was physical, and only then were we able to make any causal claims, including the claim that the mental quale were caused by the physical objects. This causal hypothesis was intelligible

[3] Personal communication, c. 1971.

only after the physical objects were there, were 'posited'. It is clear that on any such account the inference to the existence of physical objects is not deductive, so the legitimacy of the inductive inference is a prerequisite for the rationality of the whole enterprise. And one cannot depend on reliability being ensured by causation, as this tactic is available only after the 'positing' of physical objects has succeeded—or failed.

It is clear from the above that the legitimacy of induction is crucial to any story about our entitlement to knowledge of such 'external' and independent physical objects, given that inductive inference forms such a large part of our entitlement to any knowledge going beyond perceptual experience. For this reason *Probability and Evidence* has to be seen as an integral part of the constructive project of *The Origins* and *The Central Questions*: without it, the innovative aspects of these books would have been crucially incomplete.

The aforementioned turn to causation to underwrite inductive inference is not a move Ayer would have approved of anyway; the influence of Hume is apparent everywhere in the formation of his philosophical outlook, and his Humean view of causation reduced cause to a (specific kind of) regularity. As a consequence, the reliability of any supposed causal link, the right to depend on the regularity being persistent, would have to be established prior to any dependence on it to ground inductive inference, so the very attribution of a causal link presupposed inductive inferences (Foster 1985:239–263 contains a critical examination of Ayer's views on causality).

The threat of circularity was endemic to all attempted justifications of induction, or so Ayer argued. As he diagnosed it, the central problem with any justification of these inferences was that they presumed the uniformity of nature, an idea Ayer put in terms of assuming that the future will, in relevant respects, resemble the past (1956:72). Inductive inferences are not necessarily about the future, however, so the assumption is better put in terms of the unobserved resembling, in relevant respects, the observed. Given that any 'principle' of uniformity of nature was not demonstrable, relying on such a principle either was unfounded or presupposed what it was meant to justify—the rationality of induction. A similar argument applied to any other principles that may have been thought to supply the missing ingredient, such as an appeal to laws of nature or to 'fair' sampling procedures. Laws of nature were not demonstrably knowable, and presuming the fairness of a sample was, in this context, to beg the question at hand.

In *Probability and Evidence* Ayer examined the problem of induction in greater detail, particularly in relation to attempts to make the problem tractable by appeal to notions of probability. The first section of the book deals in part with this problematic, before proceeding to look at the qualitative concept of confirmation. The second section of *Probability and Evidence* is something of an oddity, as it examines the inductive logic proposed by the economist Roy Harrod. It had been published previously as "Has Harrod Answered Hume?" in a Festschrift in honour of Harrod.[4] It may seem strange that Ayer was interested in an economist's views on what he took to be essentially a philosophical problem, but his interest in Harrod's work on induction (and possibly that of Keynes) stemmed from conversations with Harrod during Ayer's tenure as a young lecturer at Christ College, Oxford. Harrod was a Fellow of the same college, had been taught by the economist Maynard Keynes, and wrote on probability and the problem of induction, as had Keynes before him. (It has been argued that Keynes's own work on probability was much influenced by particular aspects of his economic theory, his interpretation of probability changing as the economic problem in view changed [see Cate and Johnson 1998, and also Cottrell 1993].) In the final section Ayer examines a problem arising from Nelson Goodman's work on confirmation, that of the proper treatment of conditionals, especially what have come to be called counterfactual conditionals. In what follows I will briefly look at some of the topics addressed by Ayer in sections I and III, pointing readers to more recent literature on some of the issues raised.

Probability

As mentioned, Ayer's interest in probability derived from the hope that the concept could be useful in defusing the problem of induction. The intuitive idea is that inductive 'logic' could be made to work, and be seen to be rational, if one replaced the deductive entailment relation with an inductive 'probabilifying' relation. Slightly more precisely, an inductive inference is said to be legitimate if the premises render the conclusion's truth 'highly probable', or, in other words, if they highly confirm the conclusion. Rudolf Carnap was one of the leading exponents of this line of thought, developing it in increasing detail throughout his career

---

[4] The Festschrift is *Induction, Growth, and Trade*, edited by W. A. Eltis, M. F. Scott, and J. N. Wolfe (London: Clarendon Press, 1970).

(Carnap 1950, 1971, 1980). What is significant here is that Carnap's use
of probabilities was tied to what is called the 'logical' interpretation of
probability, and it is this to which Ayer objected. His major objection,
that probability thus construed could not be a 'guide to life', is indicative
of Ayer's belief that any legitimation of induction should have a practical
application. That is, it should legitimate our best inductive practice; if it
couldn't do that, then legitimating induction would be a rather futile
theoretical exercise, with no bearing on whether any of our actual, or
achievable, inductive inferences has any rational warrant.

This 'requirement of practicality' shows up in an earlier article attacking the idea that the logical conception of probability could be a useful
guide to the future (Ayer 1957). Given a proposition, a, that a horse is
going to win the race, and various sources of evidence, h_1, h_2, h_3-h_n,
one can estimate the (conditional) probability of a given h_1 to be p_1,
given h_2 to be p_2, and so on. One can also estimate the probability of a
given all of h_1-h_n. Call this probability p_n, it being the probability of a
given all of the evidence available to the person wishing to place a bet
on the horse. Which of these probabilities, asks Ayer, would it be rational
for this person to bet on? Common sense dictates that p_n is the best
estimate, but, Ayer argues, on the logical conception of probability, all
of the estimates p_1-p_n are logically true, and so it is impossible to single
out one as being 'better' than any of the others. None can be preferred
over any of the others as supplying a better reason to make a bet.

Again, common sense (and Carnap) say that a probability based on
'total' evidence is what is needed. But why, asks Ayer, do we have to take
into account total evidence? Given that all of the different estimates are
logically true, there can be nothing wrong in relying on one rather than
another. Saying that if one takes into account all the available evidence,
one is more likely to be right is equivalent to saying that the hypothesis
that 'those with total evidence' are more often right has a certain probability, and this gets us no further forward. (Ayer revisits, and restates, this
objection in his "Replies" in Macdonald 1979, where he is responding
to John MacKie's claim to have rebutted the objection [MacKie 1979].
An earlier discussion of Ayer's scepticism about the use of any 'principle
of total evidence' is contained in Good 1967).

Ayer took this result to show that the logical interpretation of probability could not help if estimates of probability were to function as a
'guide to life'. He elaborated on this rejection in the more extended
treatment of probability in *Probability and Evidence*. Here he notes
other distinctive characteristics of Carnap's approach that make it an

unattractive solution to the problems of induction and confirmation. One aspect is that the syntactic approach favoured by Carnap makes the probability estimates language-relative; the initial choice of predicates will determine the probabilities of the states described by those predicates. Change the language (by changing the predicates), and one will arrive at different estimates of probability. (For a recent discussion of such language-relativity in Carnap's system, see Fitelson 2005).

This problem reappears in slightly different guise in Ayer's discussion of the frequency interpretation of probability, where he notes that the probability of an event will change according to the reference class to which it is assigned; the size of a reference class is determined by the number of events sharing relevant properties. Again, the intuitive view is that one should estimate the probability of an event using all the possible information, thus determining the 'narrowest' reference class—one that includes all relevant properties. Ayer objects that this view is not validated by anything provided by the frequency theory, which by itself cannot entitle one to 'privilege' one reference class over another.

Ayer also advises against yielding to the temptation to conclude anything about *individuals* from the information provided by a probability, where this is interpreted as a frequency. This mistake (of moving to a conclusion about an individual on the basis of information about frequencies) has had ramifications for legal processes in the United States in a case where a judge was required to sentence a drug smuggler. The sentence depended on the total amount of drugs smuggled, but though the smuggler, a Nigerian, Charles Shonubi, had been found on one occasion to have 427.4 grams of heroin, he had also smuggled drugs on seven previous occasions. The problem was how to estimate the total amount of drugs smuggled. The prosecution's argument for finding there to be a probability of .99 that Shonubi had smuggled a total of about 2,500 grams depended on statistical evidence. Records showed that other Nigerians using the same drug-smuggling technique and going through the same airport, between the times Shonubi was active (all of these features determining the reference class in which to place Shonubi), would have smuggled a total, for seven occasions, of 2,090 grams. This added to the 427 grams on the last venture gave the prosecution the 2,500-gram estimate. The prosecution's case required the judge to draw a conclusion about the individual, Shonubi, on the basis of what other individuals had done, which he declined to do; Ayer would have applauded the decision, though it was not well received by commentators. (See the discussion of the case in Colyvan, Regan, and Ferson 2003).

Two other themes emerge from the examination of Carnap's system, and both recur in later discussions. The first has to do with the ubiquity of assumptions that smuggle into the argument the principle of the uniformity of nature, or methods of proceeding that assume such a principle. Ayer later declared that there was no tenable form of this principle that would validate inductive reasoning, though all attempts to justify inductive reasoning in the end relied on it, or something close to it (see Ayer, "Replies," in Macdonald 1979:300). The second, related theme concerns the starting point for assigning probabilities. For Carnap the state descriptions (the result of applying predicates to individuals) have to be assigned unconditional, or *a priori*, probabilities, and the immediate question arises as to what justifies any such assignment. This problem recurs in Ayer's discussion of Keynes's view, where an initial assignment of probability has to be assumed, there again being a question as to how these prior assignments are justified. Ayer's challenge to the role of probability (as interpreted by Carnap) in application to empirical matters has been lucidly discussed by Bela Juhos (1969). Foster (1985:198–227) has an in-depth examination of Ayer's views on induction (especially in relation to Ayer's take on Hume's argument), *a priori* probability, and confirmation.

Recent Work on Probability

Philosophical work on the interpretations of probability since 1971 has spawned an enormous number of books and articles, many of them very technical. What follows is inevitably a somewhat idiosyncratic selection. A comprehensive survey that appeared shortly after *Probability and Evidence* is Terrence Fine's *Theories of Probability* (1973). A more recent book-length philosophical treatment is that of Donald Gillies, *Philosophical Theories of Probability* (2000). Gillies reviews the theories of Carnap and Keynes, the frequency theory of Ludwig von Mises, and the two propensity theories put forward by Karl Popper, the first holding propensities to be responsible for the observed frequencies in repeatable experiments, the second ('single-case propensity theory') regarding a propensity as that which produces a specific result on a particular occasion. Popper thought that propensities were needed in order to make sense of some results in quantum mechanics (see Popper 1957, 1959).

Gillies eschews single-case propensities before defending something like the first of the two propensity theories, which maintains that probability is a *disposition to produce* the relative frequencies mentioned by von Mises; this disposition can explain why a type of event occurs with a particular frequency. (Ayer rejected Popper's propensity interpretation in part because it was too 'metaphysical': either the propensity was a physical force, in which case it wasn't a probability, or it was something else—but what?[5])

The best guides to the later literature are recent encyclopedia articles, especially Alan Hajek's "Interpretations of Probability" in *The Stanford Encyclopedia of Philosophy* (Hajek 2003). Hajek outlines the probability calculus before discussing the various interpretations, including Carnap's logical interpretation, von Mises's frequency interpretation, the propensity account, and subjective probability. This contains a guide to the literature, both recent and not so recent. Also in the *Stanford Encyclopedia* is the more technical "Inductive Logic" by James Hawthorne (2004), containing useful material on prior and *a posteriori* probabilities, and Bayes's theorems. This entry also has valuable links to other electronic resources. There are also two articles in *Philosophy of Science: An Encyclopedia,* one by Fitelson, Hajek, and Hall ("Probability") and the other by Fitelson ("Inductive Logic"), relevant to Ayer's concerns (see Pfeifer and Sarkar 2005). The latter is especially useful in linking induction to probability, as is the recent reissue of *Choice and Chance* by Brian Skyrms (1999). Ian Hacking has a recent well-reviewed textbook on probability and inductive logic, where he introduces the basic ideas, then discusses conditional probability, Bayes's rules, and different interpretations of probability (Hacking 2001).

Theory and Evidence by Clark Glymour (1980) has a chapter on empiricist views on confirmation, and also a chapter on Bayesianism. Bayesian views, which mostly interpret probabilities as subjective, as degrees of belief, were not much discussed by Ayer; my suspicion is that he rejected them out of hand because they require prior probabilities to be assigned before they can operate on incoming data. For the issue concerning Ayer, the legitimacy of inductive inference, the assignment of these prior probabilities would be seen as begging the question. For a history

---

[5] Personal communication, c. 1970. Ayer also rejected the propensity theory because he thought that in some cases (such as a horse race) there was no single item to which the propensity attached—it was 'spread out' over the field (Ayer 1957). Popper later revised the propensity theory to make the propensity attach to the whole preceding state of the universe.

of early twentieth-century Bayesian and anti-Bayesian controversies, see David Howie's account of the debates between (the Bayesian) Harold Jeffreys and (the non-Bayesian) R.A. Fisher (Howie 2002). For more recent discussions and defences of Bayesianism, see Earman (1992) and Howson and Urbach (2002). David Lewis's *Philosophical Papers Volume II* contains some of his important work on probability, including the influential "The Subjectivist's Guide to Objective Chance" (Lewis 1980, 1986). Two recent articles that defend probability as a 'guide to life' are Papineau and Beebe (1997) and Hajek (2003).

Confirmation

The first part of *Probability and Evidence* ends with a discussion of the problem of confirmation in its most fundamental form. In assessments of the probability of any hypothesis, one takes for granted both that one has evidence of a certain kind for the hypothesis *and* that the hypothesis is one for which evidence can be given. Carl Hempel, in his papers on confirmation, made the brave attempt to be precise about what counts as evidence (see Hempel 1965). His notorious 'paradox of the ravens' led him, on the basis of the plausible assumptions outlined by Ayer, to conclude that the claim 'All ravens are black' could be evidentially supported by intuitively irrelevant items, such as red handkerchiefs. Ayer explores various attempts to avoid the counterintuitive conclusion, including one that appeals to Nelson Goodman's discussion of the 'new riddle of induction' (Goodman 1955; Foster 1985:217–227 contains a discussion of Ayer's treatment of Hempel's paradox and of Goodman's 'new riddle' of induction). Here the focus turns from considerations of what can count as confirming evidence for a hypothesis to an examination of what makes any hypothesis suitable for being confirmed. Goodman claimed that some hypotheses, such as "All emeralds are grue", are not apt for confirmation, where "grue" is a predicate true of any emerald if it has been examined and is green, and true of those emeralds that have not been examined and are blue.[6] It is clear that "all emeralds are green" and "all emeralds are grue" attribute a different colour to unobserved emeralds, but both generalisations would appear to be confirmed to an equal degree by the evidence, given that the evidence consists in all the observed emeralds (and presuming these are green). Goodman held that

---

[6] The 'new riddle' has been formulated in many ways; I am expressing the simplest formulation here. For Goodman's formulation, see Nelson Goodman, *Fact, Fiction, and Forecast*, 2nd ed. (New York: Bobbs-Merrill, 1965).

predicates such as "grue" were not projectible and could not be used in generalisations (such as "All emeralds are grue") for predictive purposes; so generalisations employing such nonprojectible predicates were not apt for being confirmed, so could not be confirmed by anything. Green emeralds hence are irrelevant.

The application of this notion of 'nonprojectibility' to the raven paradox involves declaring the predicates 'nonblack' and 'nonraven' to be nonprojectible, and so preventing their use in confirmable generalisations. The principle behind this ploy states that the logical complement of a projectible predicate is nonprojectible. If 'black' is projectible, then 'nonblack' is ruled out—if the principle is sound. I leave to the reader the assessment of whether Ayer's rejection of this manoeuvre is warranted, but do want to draw attention to the typically Humean manner in which Ayer ends this section. He in effect concedes that procedures can only be warranted as rational by a standard that will presume the rationality of the procedures in question, such circularity being unavoidable. Any attempt to convince someone who uses different standards of rationality of the error of their ways would end up simply begging the question, as we would be assuming the correctness of our own standards in any such argument. The best we can do is go on as before, hoping that previous successes will extend to future uses of such rational procedures.

Recent Work on Confirmation

Again, the literature spawned by Hempel's paradoxes of confirmation and Goodman's new riddle of induction is vast, so I will point the reader both to work that makes a significant contribution and to surveys that contain further guides to the literature.

The classic presentation of the paradoxes is found in Hempel (1965), and the new riddle of induction is first formulated in Goodman (1955). An early examination of Hempel's paradox is to be found in Schlesinger (1974). One of the clearest, and best, discussions of Goodman's new riddle of induction is provided by Skyrms (1999), who discusses it conjunction with material on probability and the traditional problem of induction. There is a large volume of essays edited by Douglas Stalker evaluating the various solutions proposed in the forty-odd years after Goodman's book appeared (Stalker, ed. 1994). More recently a solution has been put forward by Peter Godfrey-Smith, who argues that the problem, or one form of the problem, can be defused using meth-

ods already available to scientists (Godfrey-Smith 2003). There is also a
discussion of both Hempel's paradox and Goodman's 'grue' hypothesis
from the point of view of learning theory in Oliver Schulte's survey
"Formal Learning Theory" in *The Stanford Encyclopedia of Philosophy*
(Schulte 2005).

Conditionals

In the final section of *Probability and Evidence* Ayer switches from examining aspects of the problem of induction to looking at another problem
analysed by Nelson Goodman, the difficult topic of conditionals. The
problem is that the material conditional "if p then q" is true whenever
the antecedent is false or the consequent true; this yields the counter-
intuitive result that all conditionals with false antecedents ('counterfactuals') are true. Ayer broadens the concern to all those conditionals
that behave in a nontruth-functional way, and relies on distinguishing
between the truth of a conditional and its acceptability. Nontruth-functional conditionals are acceptable, he says, when we have some evidence
for the implied connection between antecedent and consequent; this
can be as weak as the rather skimpy evidence supporting a *tendency* for
the antecedent state of affairs to be followed by the consequent state of
affairs. Such fragile evidence will be sufficient for the acceptance of the
conditional linking the two if there is no stronger evidence supporting
the existence of a countertendency.

Subsequent work on conditionals and counterfactuals has been extensive, and again I will mention only works that have had a large impact, or recent work that has the virtue of summarising the existing state
of play. A major work published shortly after *Probability and Evidence* is
David Lewis's *Counterfactuals* (Lewis 1973), which analyses counterfactuals using the apparatus of possible worlds. A similar approach is adopted by Robert Stalnaker (1984), though he applies the possible-world
analysis to all types of conditionals. Also seminal is Ernest Adams's *The
Logic of Conditionals* (1975), an account of conditionals that embraces a
probabilistic approach, as Igal Kvart does in his comprehensive treatment in *A Theory of Counterfactuals* (1986). Jackson (1987) separates out
the question of the truth-conditions for conditionals from that of their
assertability-conditions, claiming that the latter form part of the meaning of the conditional. David Sanford's *If P Then Q* (1989) contains a
history of work on conditionals prior to 1989, together with arguments
for his own preferred treatment. In *Logical Forms* (1991), Mark Sainsbury

has an excellent general discussion of problems of truth-functionality in English, with a separate chapter on conditionals.

Another very useful overview of recent work on conditionals, particularly indicative conditionals, is provided by Dorothy Edgington's 'state of the art' survey article (Edgington 1995). Edgington provides another overview of the territory in the *Stanford Enyclopedia* entry "Conditionals" (Edgington 2001). There are also valuable collections of articles on conditionals, such as Jackson (1991) and the volume edited by Harper, Stalnaker, and Pearce (1981), both of which contain some of the seminal articles. More recent book-length studies of conditionals are William Lycan's *Real Conditionals* (2001) and Jonathan Bennett's important *A Philosophical Guide to Conditionals* (2003), which is illuminatingly reviewed by Lycan in *Mind* (2005).

BIBLIOGRAPHY

Selection of A. J. Ayer's Publications

1936. *Language, Truth, and Logic*, 2nd ed. London: Gollancz, 1946.
1940. *The Foundations of Empirical Knowledge*. London: Macmillan.
1956. *The Problem of Knowledge*. London: Macmillan.
1957. "The Conception of Probability as a Logical Relation." In S. Korner, ed., *Observation and Interpretation in the Philosophy of Physics*. New York: Dover.
1963. *The Concept of a Person and Other Essays*. London: Macmillan.
1968. *The Origins of Pragmatism*. London: Macmillan.
1969. *Metaphysics and Common Sense*. London: Macmillan.
1971. *Russell and Moore: The Analytical Heritage*. London: Macmillan.
1972. *Probability and Evidence*. London: Macmillan.
1973. *The Central Questions of Philosophy*. London: Weidenfeld.
1979. "Replies." In G. Macdonald, ed., *Perception and Identity*. London: Macmillan.
1980. *Hume*. Oxford: Oxford University Press.
1982. *Philosophy in the Twentieth Century*. London: Weidenfeld.
1984. *Freedom and Morality and Other Essays*. Oxford: Clarendon Press.
1986. *Ludwig Wittgenstein*. London: Penguin.

Ayer published two volumes of autobiography:
1977. *Part of My Life*. London: Collins.
1984. *More of My Life*. London: Collins.

A very informative, and philosophically informed, biography of Ayer has been written by Ben Rogers:

Rogers, Ben. *A. J. Ayer: A Life.* London: Grove Press, 2002.

Books and Collected Essays on the Philosophy of A. J. Ayer

Foster, J. *A. J. Ayer.* London: Routledge, 1985.

Griffiths, A. P. *A. J. Ayer Memorial Essays.* Cambridge: Cambridge University Press, 1991.

Hahn, L. E. *The Philosophy of A. J. Ayer.* Chicago and LaSalle, IL: Open Court, 1992.

Honderich, T. *Essays on A. J. Ayer.* Cambridge: Cambridge University Press, 1991.

Macdonald, G. F., ed. *Perception and Identity.* London: Macmillan, 1979.

Macdonald, Graham and Crispin Wright, eds. *Fact, Science, and Morality.* Oxford: Blackwell, 1986.

Martin, R. *On Ayer.* New York: Wadsworth, 2000.

Some Recent Work on Probability and Evidence

Cate, Thomas and L. E. Johnson. "The Theory of Probability: A Key Element in Keynes' Revolution." *International Advances in Economic Research* 4, no. 4 (1998): 324–335.

Colyvan, Mark, Helen Regan, and Scott Ferson. "Is It a Crime to Belong to a Reference Class?" In Henry Kyburg and Miriam Thalos, *Probability Is the Very Guide of Life* (Chicago and La Salle, IL: Open Court, 2003).

Cottrell, A. "Keynes' Theory of Probability and Its Relevance to His Economics." *Economics and Philosophy* 9 (1993): 25–51.

Earman, John. *Bayes or Bust.* Cambridge, MA: MIT Press, 1992.

Fine, Terrence. *Theories of Probability.* New York: Academic Press, 1973.

Fitelson, Branden. "Inductive Logic." In J. Pfeifer and S. Sarkar, eds., *Philosophy of Science: An Encyclopedia* (London: Routledge, 2005).

Fitelson, B., A. Hajek, and N. Hall. "Probability." In J. Pfeifer and S. Sarkar, eds., *Philosophy of Science: An Encyclopedia* (London: Routledge, 2005).

Gillies, Donald. *Philosophical Theories of Probability.* London: Routledge, 2000.

Glymour, Clark. *Theory and Evidence.* Princeton: Princeton University Press, 1980.

Godfrey-Smith, Peter. "Goodman's Problem and Scientific Methodology." *The Journal of Philosophy* C, no. 11 (2003): 573–590.

xxiv

GRAHAM MACDONALD

Good, I. J. "On the Principle of Total Evidence." *British Journal for the Philosophy of Science* 17 (1967): 319–321.

Goodman, Nelson. *Fact, Fiction, and Forecast.* 2nd ed. (New York: Bobbs-Merrill, 1965).

Hacking, Ian. *An Introduction to Probability and Inductive Logic.* Cambridge: Cambridge University Press, 2001.

Hajek, Alan. "Conditional Probability Is the Very Guide of Life." In Henry Kyburg and Miriam Thalos, *Probability Is the Very Guide of Life* (Chicago and La Salle, IL: Open Court, 2003), 183–204.

———. "Interpretations of Probability." In *The Stanford Encyclopedia of Philosophy* (Summer 2003 edition), Edward N. Zalta, ed., http://plato.stanford.edu/archives/sum2003/entries/probability-interpret/.

Hawthorne, James. "Inductive Logic." In *The Stanford Encyclopedia of Philosophy* (Fall 2004 edition), Edward N. Zalta, ed., http://plato.stanford.edu/archives/fall2004/entries/logic-inductive/.

Hempel, Carl. *Aspects of Scientific Explanation.* New York: The Free Press, 1965.

Howie, David. *Interpreting Probability: Controversies and Developments in the Early Twentieth Century.* Cambridge: Cambridge University Press, 2002.

Howson, C. and P. Urbach. *Scientific Reasoning: The Bayesian Approach.* La Salle, IL: Open Court, 1989.

Juhos, Bela. "Logical and Empirical Probability." In *Logique Et Analyse* XII, no. 47 (September 1969): 277–282.

Korner, S., ed. *Observation and Interpretation in the Philosophy of Physics.* New York: Dover, 1957.

Kyburg, Henry and Miriam Thalos. *Probability Is the Very Guide of Life.* Chicago and La Salle, IL: Open Court, 2003.

Lewis, David. "A Subjectivist's Guide to Objective Chance." In R. C. Jeffrey, ed., *Studies in Inductive Logic and Probability,* vol. II. Berkeley: University of California Press, 1980.

———. *Philosophical Papers Volume II.* Oxford: Oxford University Press, 1986.

MacKie, J. "A Defence of Induction." In G. F. Macdonald, ed., *Perception and Identity* (London: Macmillan, 1979), 113–130.

McGrew, Timothy. "Direct Inference and the Problem of Induction." In Henry Kyburg and Miriam Thalos, *Probability Is the Very Guide of Life* (Chicago and La Salle, IL: Open Court, 2003) . (Contains a discussion of the arguments on inductive rationality in *Probability and Evidence.*)

Papineau, David and Helen Beebe. "Probability as a Guide to Life." *The Journal of Philosophy* 94 (1997): 217–243.

Pfeifer, J. and S. Sarkar, eds. *Philosophy of Science: An Encyclopedia.* London: Routledge, 2005.

Popper, Karl. "The Propensity Interpretation of Probability." *British Journal of the Philosophy of Science* 10 (1959): 25–42.

Schlesinger, G.. *Confirmation and Confirmability.* Oxford: Clarendon Press, 1974.

Schulte, Oliver. "Formal Learning Theory." In *The Stanford Encyclopedia of Philosophy* (Summer 2005 edition), Edward N. Zalta, ed., http://plato.stanford.edu/archives/sum2005/entries/learning-formal/.

Skyrms, B. *Choice and Chance.* 4th. ed. New York: Wadsworth, 1999.

Some Recent Work on Conditionals

Adams, Ernest. *The Logic of Conditionals.* Dordrecht: Reidel, 1975.

Bennett, Jonathan. *A Philosophical Guide to Conditionals.* Oxford: Clarendon Press, 2003.

Edgington, Dorothy. "On Conditionals." *Mind* 104 (1995): 235–329.

——. "Conditionals." In *The Stanford Encyclopedia of Philosophy* (Fall 2001 edition), Edward N. Zalta, ed., http://plato.stanford.edu/archives/fall2001/entries/conditionals/.

Harper, W. L., R. Stalnaker, and G. Pearce, eds. *Ifs: Conditionals, Belief, Decision, Chance, Time.* Dordrecht: Reidel, 1980.

Jackson, Frank. *Conditionals.* Oxford: Blackwell, 1980.

Jackson, Frank, ed. *Conditionals.* Oxford: Oxford University Press, 1991.

Kvart, Igal. *A Theory of Counterfactuals.* Indianapolis: Hackett, 1986.

Levi, Isaac. *For the Sake of Argument: Ramsey Test Conditionals, Inductive Reasoning, and Nonmonotonic Reasoning.* Cambridge: Cambridge University Press, 1996.

Lewis, David. *Counterfactuals.* Oxford: Basil Blackwell, 1973.

Lycan, William. *Real Conditionals.* Oxford: Oxford University Press, 2001.

——. "Review of *A Philosophical Guide to Conditionals* by Jonathan Bennett." *Mind* 114 (2005):116–119.

Sanford, David. *If P Then Q: Conditionals and the Foundations of Reasoning.* London: Routledge, 1989.

Stalnaker, R. *Inquiry.* Cambridge, MA: MIT Press, 1984.

Preface

The three-part essay from which this book takes its title is based on a series of three lectures which I delivered at Columbia University on 7, 8 and 9 April 1970. They constituted the second series of John Dewey Lectures, of which the first had been given two years previously by Professor W.V. Quine and resulted in his book *Ontological Relativity*.

The title 'Probability and Evidence' was also that of a series of four lectures which I delivered, as the Shearman Memorial Lectures, at University College, London, in May 1964. These lectures already contained some of the main ideas that I tried to develop in the John Dewey Lectures, but I hope that I improved on them in various respects.

The lectures on 'The Legacy of Hume' and '*A priori* Probability and the Frequency Theory' are reproduced here substantially in the form in which they were delivered, but the third lecture, on 'The Problem of Confirmation', has been considerably expanded. This expansion mainly consists in a much fuller treatment of Hempel's paradox. In exploring this question I have profited greatly by reading Professor Israel Scheffler's book *The Anatomy of Inquiry*. My general debt to Professor Nelson Goodman will be obvious to anyone who is familiar with his book *Fact, Fiction and Forecast*.

In order to bring this book up to a respectable size, I have included two further essays in it. The one entitled 'Has Harrod Answered Hume?', from which I have made a few excisions to avoid repetition, was originally written as a contribution to a volume of essays in honour of Sir Roy Harrod which was published in 1970 by the Oxford University Press under the title of *Induction, Growth and Trade*. I have to thank the Oxford University Press for permission to reprint it. The remaining essay, on 'The Problem of Conditionals', has not been previously published. I wrote it subsequently to the John Dewey Lectures in an attempt to fill in some of the gaps which they seemed to me to leave; but I am well aware that many gaps still remain to be filled.

Finally, I wish to thank Mrs Guida Crowley for the work she has done, first in typing the lectures, and then in helping to prepare the book for the press.

A. J. AYER

New College,
Oxford
15 May 1971

I Probability and Evidence

1 The Legacy of Hume

A. HUME'S FORMULATION OF THE PROBLEM OF INDUCTION

A rational man is one who makes a proper use of reason: and this implies, among other things, that he correctly estimates the strength of evidence. In many instances, the result will be that he is able to vindicate his assertions by adducing other propositions which support them. But what is it for one proposition to support another? In the most favourable case, the premises of an argument entail its conclusion, so that if they are true the conclusion also must be true. It would seem, however, that not all our reasoning takes the form of deductive inference. In many cases, and most conspicuously when we base an unrestricted generalisation on a limited set of data, we appear to run beyond our evidence: that is, we appear not to have a logical guarantee that even if our premisses are true, they convey their truth to the conclusion. But then what sort of inference are we making, and how can it be justified? These questions have not proved easy to answer, and their difficulty creates what philosophers call the problem of induction.

The attention which has been paid to this problem is primarily due to the work of David Hume; and it is worth taking some trouble to restate Hume's argument since, for all its essential simplicity, it has often been misunderstood. Hume starts from the assumption that we can have reason to believe in the truth of any proposition concerning an empirical matter of fact only in so far as we are able to connect the state of affairs which it describes with something that we now perceive or remember. Let us assign the neutral term 'data' to what one perceives at a given moment or what one then remembers having previously perceived, leaving aside the question what these data are. On any theory of perception, their range will be very limited. Then Hume maintains that one will have reason to believe in the existence of anything which is not a datum, at the time in question, only if one has reason to believe that it is connected with one's data in a lawlike fashion. He puts this rather misleadingly by saying that 'all reasonings concerning matters of fact seem to be founded on the relation of cause and effect'.[1] He then raises the question whether our belief in the existence of these lawlike connections can ever be rationally justified, and he offers a proof that it cannot.

This proof may be set out in nine stages, as follows:

(i) An inference from one matter of fact to another is never demonstrative. This is not to say that when the inference is fully set out the conclusion does not follow validly from the premises – there is no question but that 'q' does follow from 'p' and 'if p then q' – but rather that what one may call the guiding principle of the inference, the proposition 'if p then q', when based on a supposed factual connection between the events referred to by 'p' and 'q', is always an empirical proposition, and as such can be denied without contradiction. Hume's way of putting this, or one of his many ways of putting this, is to say that 'knowledge of the relation of cause and effect is not, in any instance, attained by reasonings *a priori*, but arises entirely from experience'.[2]

(ii) There is no such thing as a synthetic necessary connection between events. These are not, of course, the terms in which Hume puts it, but this is what it comes to. No matter what events A and B are, if A is presented to us in some spatio-temporal relation to B, there is nothing in this situation from which we could validly infer, without the help of other premises, that events of the same type as A and B are connected in the same way on any other occasion. There is no such thing as seeing that A *must* be attended by B, and this not just because we lack the requisite power of vision but because there is nothing of this sort to be seen. No sense can be given to a 'must' of this type.

(iii) So the only ground that we can have for believing, in a case where A is observed by us and B not yet observed, that B does exist in such and such a spatio-temporal relation to A is our past experience of the constant conjunction of As and Bs.

(iv) But clearly the inference from the premiss 'Events of the type A and B have invariably been found in conjunction', or to put it more shortly, 'All hitherto observed As bear the relation R to Bs', to the conclusion 'All As bear the relation R to Bs', or even to the conclusion 'This A will have the relation R to some B', is not formally valid. There is what we may call an inductive jump.

---

[1] David Hume, *An Enquiry Concerning Human Understanding*, section iv.
[2] Ibid.

(v) To make it valid an extra premiss is needed assuring us that what has held good in the past will hold good in the future. Hume's formulation of this principle in the *Treatise of Human Nature* is 'that instances of which we have had no experience, must resemble those of which we have had experience, and that the course of nature continues always uniformly the same'.[1]

(vi) But if all our reasonings about matters of fact are founded on this principle, we have no justification for them unless the principle itself is justifiable. But what justification could it have? There can be no demonstrative argument for it. It is clearly not a logical truth. In Hume's own words 'we can at least conceive a change in the course of nature; which sufficiently proves that such a change is not absolutely impossible'.[2]

(vii) Even if the principle cannot be demonstrated, perhaps we can at least show it to be probable. But a judgement of probability must have some foundation. And this foundation can lie only in our past experience. The only ground we can have for saying that it is even probable that the course of nature continues uniformly the same is that we have hitherto found this to be the case. But then we are arguing in a circle. To quote Hume again, 'probability is founded on the presumption of a resemblance betwixt those objects of which we have had experience, and those of which we have had none; and therefore it is impossible that this presumption can arise from probability'.[3]

(viii) The same objection would apply to any attempt to by-pass the general principle of the uniformity of nature and argue that inferences from one matter of fact to another, though admittedly not demonstrative, can nevertheless be shown to be probable. Again, this judgement of probability must have some foundation. But this foundation can lie only in our past experience. And so we have to assume the very principle that we are trying to by-pass, and the same objections arise.

(ix) We must, therefore, admit that since the inferences on which we base our beliefs about matters of fact are not formally valid, and since the conclusions to which they lead cannot be shown without circularity even to be probable, there is no justification for them at all. We just have the habit of making such inferences, and that is all there is to it. Logically, we ought to be complete sceptics, but in practice we shall continue to be guided by our natural beliefs. Hume is not, indeed, so inconsistent as to regard this generalisation about human behaviour as

[1] David Hume, *A Treatise of Human Nature*, book I, section vi.
[2] Ibid.   [3] Ibid.

being more warranted than any other. He just expresses his natural belief in it, and leaves the matter there.

Whatever we may think of the conclusion, this is a marvellous chain of arguments; one of the most brilliant examples of philosophical reasoning that there has ever been, and also one of the most influential. By going through it in detail, we shall be able to see, among other things, how different theories of induction arise from the attempt to challenge different stages in Hume's argument.

B. THE ASSUMPTION OF ATOMICITY

Let us begin with the first step: the proof that inferences concerning matters of fact are not demonstrative. Although this conclusion is by now fairly generally accepted, we need to proceed rather carefully in order to see exactly what point Hume is making. He himself expresses it by saying that every effect is a distinct event from the cause, and therefore cannot be discovered in it, but his speaking of causes and effects conceals the generality of his argument, and the notion of discovering one event in another also needs to be elucidated.

For this purpose I shall introduce the concept of an intrinsic description. I shall say that a description of the state of a subject S at a particular time t is intrinsic to S at t if and only if nothing follows from it with regard to the state of S at any time other than t, or with regard to the existence of any subject S' which is distinct from S, in the sense that S and S' have no common part. Then Hume's principle is the tautology that if two events are distinct in this sense, an intrinsic description of either one of them entails nothing at all about the existence or character of the other.

Since Hume's principle follows analytically from the definition which I have given of an intrinsic description, the only questions for debate are whether there can be intrinsic descriptions in my sense, and, if there can be, whether they are sufficient to describe everything that happens. The assumption that they are sufficient is the heart of Hume's atomism, which Russell and other modern empiricists inherited.

Though I am going, with some reservations, to answer yes to both these questions, I must begin by admitting that intrinsic descriptions are not at all common in ordinary usage, mainly perhaps because they are not individually very informative. For the most part we describe objects or events at least partly in terms of their actual or potential relations to other objects or events, and very often these descriptions include a tacit or explicit reference to their causal properties. Thus, a pen is something that you write with, a chair is something that can be sat on, a table is something that supports things, a match is something that produces flame when it is suitably struck, fire lights and warms, books can be read, mirrors reflect, umbrellas are waterproof. Indeed, almost all the most familiar kinds of objects are defined in this fashion. That is why it is misleading to say, as I sometimes have, that Hume proved that causality was not a logical relation. For if these words are taken, in what would be quite a natural sense, as implying that no causal statements are logically true, they state what is actually false. In any case in which a causal property enters into the definition of a kind of thing, the causal statement in which this property is ascribed to things of that kind will be logically true; and we have just seen that there are a great many statements of this sort. Not only that, but we can make their number as large as we please. In any case in which we want to make the claim that two properties are invariably associated, we can always make sure of the association by expanding the definition of the first property so that it includes the second. This is badly put, because in a strict sense it is no longer the same property that one is defining. What happens is that, starting with two linguistic expressions l and m, which stand respectively for the simple, or more likely, compound properties f and g, one continues to use the expression l but construes it as standing for the more comprehensive property fg. Very often this operation passes almost unnoticed because the changes are gradual and the connotations of the terms one is dealing with are anyhow not very clearly delimited. Consider, for example, the changes that have taken place even in the meaning of terms like 'water' as our knowledge, or what we take to be our knowledge, has increased. Do we now mean the same by the term 'water' as the Greeks meant by 'ὕδωρ'? Well, yes and no. If you look up 'water' in the *Oxford English Dictionary*, you will find under heading II, 'the substance of which the liquid "water" is one form among several; the chemical compound of two volumes of hydrogen and one of oxygen (H_2O)'. This is not a definition that the Greeks could have given. And what of heavy water which has a different chemical composition? Is it or is it not really water? Only a purblind believer in real essences could worry about such a question. Clearly, you can say what you like.

Again, if one succeeds in axiomatising a scientific theory, it is possible to give the branch of physics, or whatever it may be, the status of a geometry. This is done simply by treating the axioms of the theory not as empirical generalisations but as implicit definitions. Only what satisfies them is to count as a physical object of the sort with which the theory deals. There will seldom be much point in doing this and often some disadvantage. For instance, it might have the psychological effect of increasing resistance to revision of the theory, or the search for counter-examples might be impeded by the difficulty in formulating them; indeed, they could not be formulated as counter-examples but only as cases with which the theory might be expected to deal but did not. Nevertheless, such a course would be technically feasible.

It is, however, obvious that it is not possible to rebut Hume's argument by engaging in manœuvres of this type. The proof of this is that the process of turning empirical generalisations into analytic propositions can always be put into reverse. If a property f has been included in the connotation of an expression E which applies to the members of a class K, all that we have to do is invent an expression E' which applies just to those things that have the defining properties of K other than f and then raise the question whether everything that has the set of properties designated by E' also has f. Neither is there any escape by way of axiomatising theories and treating them as geometries. For the empirical questions which we suppress in this way all reappear when we come to ask whether the axioms are satisfied.

But while it may always be possible to dismantle a given structure of logically connected concepts, this does not mean that there are no limits to the extent that concepts can be divested of their logical attachments. There may, therefore, still be some doubt whether the rather stringent conditions which I have laid down for anything to be an intrinsic description can in fact be satisfied, or at least whether they can be satisfied to an extent which will carry the weight which Hume wishes to rest upon his argument. It is of no consequence that his first premiss is necessarily true, if it also turns out to be wholly or very nearly vacuous.

What sort of thing, then, could an intrinsic description be? Since it is not allowed to carry any causal implications, it would seem that it must be purely phenomenal: for example, a predicate of colour or shape. In fact, even predicates of this kind are not intrinsic, in the way we ordinarily employ them. For this table to be brown, it is not sufficient and indeed not necessary that it now looks brown to me, or to any other given observer; it has to be disposed to look brown to most observers under such and such conditions, perhaps also to reflect light of such and such a wave-length. There seems however, to be no reason in logic why such dispositional properties should not be replaced by occurrent ones – no reason, that is, why colour predicates should not be construed purely phenomenally.

But then there is a difficulty about the status of the individuals to which the predicates are applied. According to Hume's theory of knowledge, they can only be private, fleeting sense-impressions, which will presumably have to be identified by demonstratives. So our primitive statements would have the form 'this is f' where 'f' is an intrinsic description of the sense-datum designated by 'this'. However, when Hume comes to define the relation of cause and effect, the terms to which he ascribes it are events which are located in physical space and time, and the examples which he gives are all taken from the domain of physical objects. Since Hume officially takes our belief in the existence of physical objects to be a delusion, he is open at this point to the charge of inconsistency. How serious we think this is will depend on the view which we take of the possibility of constructing physical objects out of sense-impressions. My own view, for which I cannot argue here, is that the traditional phenomenalist programme of exhibiting statements about physical objects as shorthand for statements about sense-data cannot be carried through, but that it is possible and legitimate to start with neutral sense-qualia, and then represent statements about physical objects as elements in a secondary system which functions as a theory with respect to the primary system of sense-qualia.[1] Then, Hume's point can be put by saying that statements in the secondary system are not deducible from statements in the primary system and that these in their turn are not deducible from one another.

If I am wrong about this, and we are bound to take physical objects or events as our primary particulars, or if, as I am in fact inclined to hold, the possibility of starting with sense-qualia depends on our being able to locate them in a spatio-temporal system, then the conditions for the introduction of intrinsic descriptions will not be perfectly satisfied. Our primitive statements will be to some extent proleptic, or at the very least, the identification of an individual will presuppose that there are others to which it stands in some spatio-temporal relation. But the effect of this will be much less serious than one might imagine. For even

[1] See *The Origins of Pragmatism*, part ii, section 3, and *Russell and Moore: The Analytical Heritage*, chaps 3 and 9.

though our description of any observational datum may necessarily carry us beyond it, still the only thing to which we shall be logically committed will be the theory which is presupposed by our method of recording our observations. And while this theory cannot be questioned from a standpoint which presupposes it, it can be found insufficient in the sense that we may make observations which it proves ill-equipped to handle, and then we may decide to discard or modify it, and to characterise what we perceive in a more or less radically different fashion. Moreover, even within the framework of a given theory, it is possible to question the standard interpretation of any particular observation that one may make. If I am employing a primary system in which physical objects are the only individuals, then I cannot fail to make the presupposition that some physical objects exist; but I can still consistently raise the question whether *this* is a physical object of the sort I take it to be, or even whether *this* is a physical object of any kind at all; and I can devise a means of describing the situation which will leave these questions open. So even if it be granted, for the sake of argument, that I cannot characterise any individual x without relating it to some other individual, I am not bound to characterise it by relating it to any specific other individual y. Thus Hume's tautology that if a description of an object or event x does not include any reference to another object or event y, then from a statement which asserts the existence of x under that description, nothing whatsoever can be deduced about the existence or non-existence of y, is strong enough to serve as the first step in his argument.

C. NO PLACE FOR NATURAL NECESSITY

I pass now to the second step, the contention that there is no synthetic relation of necessitation. This is a more difficult question to discuss because it is not at all clear what is being denied. It is not clear what those who believe in synthetic relations of necessity take them to be. But I think that we can make sense of this issue, and also endow it with some interest, if we understand Hume to be denying that apart from rela-tions of comparison there could be anything more than spatio-temporal relations between events. This may sound paradoxical at first hearing, but when we consider what it comes to we shall see that it is at least very plausible. It amounts in fact to the claim, which is scientifically most respectable, that everything that happens in the world can be represented in terms of variations of scenery in a four-dimensional spatio-temporal continuum. So if one object, as we say, imparts motion to another, this consists in there being a change in their positions relatively to each other and to other objects: if one object communicates heat to another, there is a change of temperature within a given region, correlated perhaps with other changes in other regions. And in general all change can be represented in terms of the acquisition or loss of certain properties by the same or different regions. If we bring in objects to carry the qualities and to occupy the regions, we shall have to distinguish between different types of change, but the same principle still holds.

To show how this works in more detail, we may take a simple illustration of Hume's, that of a cannon in billiards. We think of the cue striking the billiard ball, and so imparting motion to it, of the ball hitting a second ball, and rebounding on to a third. All these forceful expressions, 'striking', 'hitting', 'imparting motion', 'rebounding', are expressions of causal agency, and their use in this context is perfectly correct. But everything that actually happens can be wholly described in purely undramatic spatio-temporal terms. First there is a period of time in which there is a concordant movement, that is, a relative change of place, of the player's body and the cue, then an instant at which the cue and the cue ball are in contact, or in other words, share a common spatial boundary, then a period in which this ball is in motion relatively to the other balls, the sides of the table, and so forth, then an instant at which the first ball is in contact with the second, then a period in which both balls are in motion, then an instant at which the first is in contact with the third. However long this story is continued, there is nothing in the situation that calls for anything more than the identification of various objects and the specification of their changing spatio-temporal relations. In particular there is nothing of which such terms as 'power', 'force', 'energy', 'agency' could be names. As Hume would put it, there are no impressions from which any such ideas could be derived.

Of course I am not maintaining that, when we say that the impact of the first ball causes the second ball to move, we are saying no more than that there is first a period in which the first ball is in motion and the second at rest, relatively to the sides of the table, then an instant at which they share a common spatial boundary, and then a period in which the balls are both in motion, relatively to the sides of the table and each other. Exactly what more we are saying is not easy to specify, except that the answer has to do with the fact that the use of causal expressions brings in an element of greater generality.[1] We are in some way implying that this situation is typical; that this is what normally happens or would happen in these circumstances. But the point is that the other situations with which this one is linked are of the same character. The only relations involved in them are spatio-temporal.

I have no doubt that the view which I am here attributing to Hume is right, but even if it were not it would leave his argument unimpaired. For let us suppose, *per impossibile*, that there is something in a situation of this kind for which a word like 'agency' could be a name. Let us symbolise this by the relational symbol R. Then what occurs, on this hypothesis, is not just that the first ball, A, is for an instant in contact with the second, B, after which B moves, but that A acquires the relation R to B. But this gratuitous complication achieves nothing at all. For if there is any doubt whether on another occasion, or on all other occasions, in which A is brought into contact with B in the same manner, the same result will ensue, there can be no less doubt both whether, on another occasion, A will stand in the relation R to B, when brought into contact with it in the same manner, and also whether, on another occasion, As having the relation R to B will have the same result; for example, whether, instead of imparting motion to B, it will not immobilise it or annihilate it or turn it into a swan. But, it may be said, this second question is taken care of by the nature given to the relation. If the first ball stands in the relation R to the second it *must* impart motion to it. But this will not do at all. For either this supposed relation is a phenomenal property, something detectable in the situation, or it is not. If it is, its existence will enable us to draw no conclusion with regard to any event occurring at any other time than the time at which it is manifested. If it is not a phenomenal property, then to be of any use, it must comprise in its definition the clause that any terms which are related by it must behave in such and such ways. But this just brings us back to the case where causal propositions are made true by fiat and the same objections apply. In particular, no terms could then be observed to stand in this relation: we should have to wait and see what happened, to find out if the relation held; from which it follows that our uncertainty about what will happen cannot be removed by the introduction of this entirely spurious relation.

The domain in which the strongest objection may arise to the thesis that the only relations that can obtain between different objects and events are spatio-temporal is that of mental phenomena. One super-

[1] See below, pp. 132 ff.

ficial difficulty is that there is good ground for saying that mental events are not in space. But if we take this view, our thesis will need only to be slightly modified. If mental events are not eligible as terms of spatial relations, then the only relations in which they can stand to one another or to physical events are temporal. But now it will be objected that this is a ludicrous impoverishment of our mental life. Surely the essential feature of mental processes is that they are relational, and not merely in a temporal way. One is angry *with* someone *over* something, afraid *of* something, unhappy *about* something, jealous *of* someone, one wishes or hopes *for* something, one's desires and beliefs have intentional objects. In a very large measure, the significance of our actions depends not merely on the physical but also on the social context in which they are performed, and so on the multifarious and complex relations that fall under political or economic or legal or moral concepts.

All this is true. We do use this sort of language in describing human states and actions, and it may well be that it serves a function that could not be fulfilled in any other way. I myself am inclined to take the view that intentional expressions are eliminable, but I have to admit that no satisfactory method of eliminating them has yet been found. But even if they are not eliminable, I do not think it follows that our thesis is disproved. For we have to distinguish between the events which actually occur and our explanations or evaluations of them. What actually occurs when someone is in some emotional state, of jealousy, or anger, or hope or whatever, is that he has, perhaps, a characteristic feeling, and at the same time has certain thoughts – let us, for simplicity, have him thinking aloud, so that we can transform this into 'makes a certain series of noises' – and subsequently perhaps behaves in certain ways, that is, makes a certain series of movements. But surely the noises must mean something to him. Yes indeed, but in so far as their meaning something to him is not a matter of what he *would* say and do, that is, an unfulfilled conditional and so not a description of anything that actually happens, all that this comes to is his making other series of noises or other movements. The meaning which is attributed to the noises, the intentional objects which are assigned to his emotions, the social significance which is read into his behaviour belong to a secondary explanatory system. But the primary events, in terms of which alone the explanations can be cashed, stand in none but spatio-temporal or merely temporal relations to each other.

I have developed this point because I think that it is of interest, but its validity is not essential to the main argument. For even if I am wrong in what I have just been saying, and the account which I wish to give of psychological facts is too austere, it still would be vain to expect them to provide a home for a concept of synthetic necessity.

No doubt we get our idea of causation from the exercise of our wills: from the experience of having objects resist us and overcoming this resistance, from carrying out our intentions and so forth. But none of this introduces any new element into the argument. However the process of volition is analysed, whether, in a case where some intention is carried out, the initial state is supposed to consist in a mental event of some kind, a feeling of desire, a representation to oneself of the end to be attained, or in a contraction of the muscles, an incipient physical movement, or some combination of the two, it must be conceivable that this initial stage should not be succeeded by the remainder of the process. That is to say, unless we beg the question by describing the initial state in a way which *entails* that the process will be completed in the orthodox fashion, it must be logically consistent with its existence that it should be immediately followed not by the intended action but by some quite different action, or by no action at all, by the agent's sudden paralysis or sudden death.

There is no need to belabour this point because it must be quite obvious that no alleged experience of necessity can achieve what is demanded of it. What is demanded is an assurance that a connection (be it merely a concomitance, which I think is all it ever could be, or some stronger relation), which is observed in one instance, should also hold between objects of the same sort in other instances, or even in all instances. But how could an experience of what goes on in this instance afford us any assurance whatever about what will occur in other instances? Even if I did perceive some necessary connection between my deciding to touch this desk and my hand's moving or my muscles' tensing (and in fact I do not perceive anything of the sort – I have no idea what it would be like to), this proves nothing whatsoever about what would happen on another occasion. Of course it is another matter if you define the decision, or the intention, by the action which follows it; but we have already seen that this device is profitless, that it leaves us exactly where we were.

It is plain that if this dubious notion of synthetic necessary connection is to have any hope of serving its purpose it must be taken as entering into general propositions. It must not be something which we experience in particular instances, but rather something which we intellectually apprehend as holding between different kinds of things or events.

In old-fashioned language the claim must be that we can apprehend necessary connections between universals.

Since 'It is necessary that p' can be taken as equivalent to 'It is not possible that not-p', the supporters of this claim may be said to be crediting themselves with a primitive concept of factual possibility. This has to be distinguished from the more generally accepted notion of logical possibility. It is not suggested, for example, that there is any contradiction in denying that the planets move round the sun in elliptical orbits, but only that it is factually impossible for them to do otherwise. Thus, what are called laws of nature are held to be necessary, without being logically true.

This interpretation of what is meant by a law of nature is advanced by Professor Kneale in his book on *Probability and Induction*. He does not have much to say in its favour, beyond rebutting Hume's argument that such propositions cannot be necessary because we are able to conceive that their contradictories might be true. Kneale's answer to this is that we can conceive propositions to be true even when they are logically impossible, so that *a fortiori* the fact that we can conceive the contradictory of a law of nature to be true does not prove that this law is not necessary, in his weaker sense of the term. His example is Goldbach's conjecture that every even number greater than 2 is the sum of two primes, which has not been demonstrated as a theorem though no exception has yet been found. If one takes the usual view that the propositions of arithmetic are logically necessary, and if one accepts the law of excluded middle, one has to conclude that one or other of the two propositions represented by Goldbach's conjecture and its contradictory is logically impossible; yet they appear to be equally conceivable. It is, indeed, a weakness in any argument of this kind that we do not have any sure method of deciding what is or is not conceivable, but this is a weakness that cuts both ways. If we are content to say that the appeal to what we can conceive is not a good criterion of possibility, we are conceding enough to meet Hume's objection.

Apart from this defensive argument, the only reason that Kneale gives for construing laws of nature as what he calls Principles of Necessitation is that he finds this less objectionable than any other account of them that has yet been suggested. The alternatives which he considers are that they are factual generalisations of unrestricted scope, in which case to assert a proposition of the simple form 'All A is B' as a law will be to assert no more than that at no place or time is there an A which is not a B, that they are factual generalisations of restricted scope, so that to assert that 'All A is B' will be to assert that there is no A within such and such a spatio-temporal region that is not a B; and lastly, that they are not propositions to which a truth-value can be assigned, but rules or prescriptions: 'Whenever you come across an A you are to rely upon its being a B.' This last view has been taken by such gifted philosophers as Schlick, Ramsey and Ryle; but it seems to be open to the serious objection that there is no reason why we should obey such a rule unless we believe that as a matter of fact all As are Bs. And once this is admitted, it seems simple and more natural to construe the assertion of a universal proposition of law as expressing the factual belief.

My own view is that generalisations of law should be taken as equivalent in content to generalisations of fact, and it would appear to be most in accordance with scientific practice to allow them unrestricted scope. This is not to say that there is no distinction at all between a generalisation of law and a generalisation of fact, but, as I have explained elsewhere,[1] I think that the distinction can be accounted for in terms of the difference in our attitudes towards them. Kneale's objection to this course is that it so weakens generalisations of law that they can no longer generate unfulfilled conditionals. From the fact that all actual chimpanzees are Rhesus-negative it cannot be inferred with regard to anything which is not a chimpanzee that if it were one it too would be Rhesus-negative. But a law of nature is supposed to apply to every possible as well as to every actual instance. On the other hand, if we construe this generalisation of law as stating that it is not possible for anything to be a chimpanzee without being Rhesus-negative, the unfulfilled conditional can be derived.

This is, indeed, a strong objection. We do not yet have a satisfactory analysis of unfulfilled conditionals and it is clear that they will have to be accommodated in any adequate account of the difference between generalisations of law and generalisations of fact. Nevertheless, I am sure that we must find some other way of dealing with them than by relying on an undefined notion of factual possibility. And my reason for saying this is that I do not think that a concept of possibility, which is other than a logical concept, can be allowed to be intelligible.

This is not, of course, to deny that the word 'possible' and its cognates can legitimately be used in a causal sense, the sense in which we say that it is not possible for a horse to breed with a mule, or for a

---

[1] See 'What is a Law of Nature?', reprinted in *The Concept of a Person*, and below, pp. 129 ff.

steam-engine to run without fuel, or for a man to walk a mile in less
than two minutes. But I believe that what is meant in such cases by
saying that something is impossible is just that it is inconsistent with
some natural law, and similarly that the only way in which a reference
to possibility can arise in connection with the law itself is with regard to
the question whether it is logically compatible with some other statement of law. Thus, the difference between logical and causal possibility
does not, in my view, consist in any different use of the word 'possibility'. It is merely that when we hold a proposition to be logically impossible, we are claiming that it is incompatible with some general
proposition which is itself logically true, and when we hold one to be
causally impossible, we are claiming that it is incompatible with some
general proposition which is, not logically, but empirically true. So, if
I am right in thinking that the concept of factual possibility derives its
sense from that of natural law, there will be a vicious circle in Kneale's
proposal to analyse the concept of natural law on the basis of a concept
of necessity which is defined in terms of the concept of factual possibility.

But even if I were mistaken on this point, it would still be clear that
this appeal to natural necessity does not help us at all in dealing with the
problem which Hume raises; and to be fair to Kneale, he has not
suggested that it does. For if we are to be justified in inferring from the
fact that something X is A to its being B, in an unexamined instance,
then either it is sufficient that 'If X is an A it is a B' should actually
state or follow from a Principle of Necessitation, whether we know
this to be so or not, in which case the mere factual truth of 'All A is B'
would be sufficient, or else we have to know, or at least have good
reason to believe, that 'If X is an A it is a B' does state or follow from a
Principle of Necessitation, and how in the world is this to be accom-
plished?

By intuition? But, once questions of logical possibility are settled, no
scrutiny of concepts is going to tell us what can or can not happen. The
only relations that can obtain between concepts are logical relations;
concepts can comprise or exclude or overlap with or be logically
indifferent to one another. And if they are logically indifferent to one
another, no inspection of them could conceivably show that they were
linked by natural necessity, even if it be granted for the sake of argument that the idea of natural necessity is intelligible. Let us say that it is a
Principle of Necessitation, or a consequence of such a principle, that the
planets move around the sun in elliptical orbits or that there are three
hundred and three thousand trillion molecules in one gramme of

hydrogen. How could we discover these things by scrutinising concepts? Why should the orbits not be circular, or the number of molecules a thousand less? The only answer is that such hypotheses would not fit, or would not so well fit, the facts which these natural laws explain. Once again, we can make these laws true by definition. We can, for example, define hydrogen in such a way that it becomes logically necessary that one gramme of hydrogen contains just that number of molecules. But then how can one possibly expect to intuit that this concept applies to anything, or that if it is found to apply to one thing it will be found to apply to others that are known to have the other defining properties of hydrogen, but not this one?

Moreover, this very appeal to intuition which enables the notion of synthetic necessary connection to play any part in this discussion also makes it superfluous. For who can intuit the greater can intuit the less. If one can know by intuition that a generalisation holds for all actual and all possible instances, then presumably one can know by intuition that it holds for all actual instances, and it that event, so far as justifying any actual prediction goes, we do not need to bring in natural necessity. But of course even this more modest feat is impossible.

When I say this I am not engaging in *a priori* psychology. I am not trying to set limitations to human powers. I have not the least objection to allowing that any number of true generalisations may be discovered by intuition. I should think it exceedingly doubtful whether any discovery of this sort could be made in the void, without any experimental knowledge of the facts which the generalisation was meant to cover, but I do not want to rule out even this far-fetched idea *a priori*. The point which I wish to make is just that a belief in a factual generalisation can never be vindicated by intuition. To say 'I apprehend that this is so' is not a proof, or at any rate not a proof in itself. It only becomes something of a proof when the person who makes a claim of this sort has been found to be always or very nearly always right, that is, when the generalisations which he has picked out in this way have never yet or only very seldom been found to come to grief. But then of course we are justifying intuition by making an inductive jump, rather than going the other way around.

D. THE PRINCIPLE OF THE UNIFORMITY OF NATURE

I now pass to the third stage in Hume's argument, the point at which he contends that the only ground we can have for believing either that all As are conjoined with Bs or that a particular A is conjoined with a particular B, in a case where A has been observed but B not, is our past experience of the constant conjunction of As and Bs.

Now this proposition is fairly obviously false as it stands, especially if we bear in mind that for Hume these As and Bs must be sense-impressions. Even if we allow the assumption that our knowledge is founded on sense-impressions to be legitimate, it is clear that even our most elementary beliefs about the physical world could not be arrived at by this straightforward process of extrapolating past regularities; there has to be a good deal of imaginative supplementation, and indeed of simplification, since any pattern which is formed by anyone's actual impressions over any length of time is likely to be highly complicated. Even if we do what Hume does in practice and think of the constant conjunctions as being conjunctions between observed physical properties, or physical events, there is only a limited number of scientific generalisations that will fit into this scheme. 'All ravens are black' can be accommodated, but not 'All inherited characteristics are transmitted by the parents' genes', and not very much of current physics, where the connection between the laws and anything observable is often remote and devious.

Nevertheless, this third proposition of Hume's, though false as he states it, is right in spirit. What we need to do is generalise it and cut it loose from its genetic ties. We can agree with Professor Popper that it does not matter for our purpose how generalisations are arrived at. The question which concerns us is what makes them acceptable. And here I think we can say that the only ground there can be for accepting any generalisation is that statements which record observed facts are derivable from it. Here I use the word 'derivable' with some latitude. The derivation may be indirect, via a series of generalisations which are as it were 'closer' to the facts, as, for example, Einstein's theory of relativity is a generalisation of Newton's theory, in the sense that Newton's theory holds within it as a special case, and Newton's theory is, in its turn, a generalisation of Kepler's laws of motion, from which observed facts about planetary motion are derivable. The derivation is, indeed, always indirect, in the sense that no particular statement is directly derivable from any generalisation even of the lowest level, but only at best from the generalisation, in conjunction with some statement which is technically said to describe the initial conditions, that is, in conjunction with some other particular statement of fact. Indeed, we may conceive of low-level generalisations as serving to link particular statements with one another, enabling us, for example, to infer from 'This is a match' to 'This will ignite under such and such conditions'. If the generalisation is statistical not even this is possible. There is no valid inference from '*n* per cent of *A*s are *B*s' and 'This is an *A*' to 'This is a *B*', but only to 'Probably this is a *B*' which is not a statement of observable fact. But we can say of statistical generalisations that they square with the facts, in cases where the observed percentage coincides nearly enough with the prescribed percentage, and I include this in my loose sense of 'derivable'.

If we generalise Hume's third stage in this way, we must also generalise the fourth. The formal invalidity of inferring from 'All hitherto observed *A*s are *B*' to 'All *A*s are *B*s' can be viewed as a special case of the general invalidity of inferring from 'Such and such a generalisation squares with all hitherto observed facts' to 'Such and such a generalisation squares with all observable facts' or even to the more modest 'Such and such a particular proposition which is derivable from the generalisation is true', in the case where the proposition in question is not itself one of those on which the generalisation has been founded. I am using 'squares with' as the converse of 'derivable from'. These are just rephrasals of the tautology that an extrapolation from a certain body of evidence is not a deduction from it. There can be no gainsaying this.

So far our examination of each stage in the argument has upheld Hume. But when it comes to the fifth step I think we are obliged to part company with him. This is the stage at which he maintains that to make the inductive jump valid we need as an extra premiss the principle that 'Instances of which we have had no experience must resemble those of which we have had experience and that the course of nature continues always uniformly the same'.

The obvious objection is that a principle so general as the one that Hume advocates cannot possibly do the work that is here required of it. We cannot validate the inference from 'All observed *A*s are *B*' to 'All *A*s are *B*s' by adding as a major premiss 'Nature is uniform'. The syllogism 'Nature is uniform', 'All hitherto observed swans have been white', therefore 'All swans are white' is not to be rejected, as

John Stuart Mill maintained, just because the minor premiss turned out to be false and so landed us with a false conclusion. It was invalid all along. To make this clear let us re-express it without the token re-flexivity, that is by putting in a date instead of the word 'hitherto'. Then it becomes, let us say 'Nature is uniform'; 'All swans observed before the year n are white'; therefore, 'All swans are white', where we may substitute any date for n which makes the minor premiss true. But then, if the syllogism is valid, the discovery of black swans refutes the uniformity of nature; for if the conclusion of a valid deductive argu-ment is false one of the premisses must be false and the uniformity of nature is here the only surviving candidate. But of course such dis-coveries are not taken as refuting the uniformity of nature. They are taken only as proving that the uniformities which nature exhibits are in some respects different from what we had supposed them to be. Not 'All swans are white' but 'All swans are non-chromatic', or 'All swans are white, under such and such conditions, or black, under such and such other conditions'.

But if nothing is going to count against the principle of the uniformity of nature, it is quite useless to us. It amounts, in fact, to saying that nature is lawlike, or, more precisely, that every event comes under some true generalisation. But if no restriction is placed on the form and complexity of these generalisations, or on the character of the predicates which enter into them, this must be trivially true. Indeed, since in any case in which we have an instance of one predicate f and an instance of another predicate g, we can turn them into instances of the same predicate merely by introducing a new predicate h, which is the dis-junction of f and g, it is clear that any universe can be made to exhibit as much repetition as we please. Suppose, however, that we refrain from securing repetition by this device, and that we are transported to a universe where none of the predicates that we have at our disposal, or those that we introduce, finds more than one instance. Even in a uni-verse of this kind, every individual falls under a true generalisation. For if only one event α of type A and one event β of type B ever occurs and α stands, as it must, in some spatio-temporal relation R to β, it will be true that for all x and y, if x is an A and y is a B, x has the relation R to y. So this extreme case of non-uniformity turns out to be an extreme case of uniformity. But, it will be objected, these generalisations are not proper laws. We cannot infer that if w were an A and z were a B, w would stand in the relation R to z. But why can we not? In depicting such a random universe, we can deal with merely possible cases in any way we please; it is not going to bring us into disaccord with any facts.

It might be thought that Hume had protected himself against this sort of trivialisation by adding the rider that instances of which we have had no experience must resemble those of which we have had experience. For let us suppose that the course of nature were to change in some startling fashion. Then, by a suitable device, we could maintain the rule of uniformity *ex post facto*, but still, it might be argued, these startling new phenomena would not *resemble* those with which we are familiar. But the trouble with this is that any two things resemble one another if they share a common quality; and, as we have just seen, any two things can be made to share a common quality. We are naturally inclined to say, for example, that two black things resemble one another in colour, whereas a black and a white one do not; but in fact the black and the white one do resemble one another, they are both non-chromatic; and so with any other example. If the predicate which marks the resemblance does not occur in our language, we can always introduce it. One might try to meet this by stipulating that only the instances of absolutely simple predicates are to count as being resemblant. But apart from the difficulty of laying down rules for determining the simplicity of predicates, this is going to prove too much. If we are going to require that instances of which we have had no experience must exactly resemble those of which we have had experience, then once more the discovery of black swans, or any other novelty, is going to destroy the uniformity of nature. So in trying to avoid making our principle so lax that it forbids nothing, we make it so stringent that any change refutes it. We shall consider later on whether there are ways of modifying it so as to make it sustain a judgement of probability, but it should by now be clear that in the role which Hume assigns to it, as the major premiss of a syllogism, it is totally incompetent.

E. KANT'S ATTEMPT TO ANSWER HUME

At the risk of flogging a dead horse, I want to say a few words about Kant's employment of the principle – in his case, in the guise of a law of universal causation – because I think that it raises some points of independent interest. Put very briefly, Kant's argument in the Second Analogy of *The Critique of Pure Reason* goes as follows: (i) The tem-

poral order in which we locate events, which we take to be empirically
real, does not simply correspond to the temporal order of our impressions. Thus, if we walk round a house we see its different sides consecu-
tively but we suppose them to exist simultaneously; we assume that
we could have obtained the same information by walking round the
house in the reverse direction. On the other hand, if we watch a boat
floating downstream, we think that the series of our perceptions does
correspond to objective successive states of the boat. In this case, the
sort of reversibility that we can introduce into our perceptions of the
house is not possible: we cannot see the boat downstream before we
see it upstream; the facts do not permit it. (ii) The explanation of this
difference is that the objective temporal ordering of events is carried
out in accordance with causal laws. The causal laws which govern the
behaviour of the materials which make up undemolished houses require
that the front and back exist simultaneously. The laws which govern
the behaviour of boats require that positions which are further from the
boat's starting-point in the direction in which it is travelling should be
reached later than positions which are nearer to it. (iii) But it is, on
Kant's view, a condition of the possibility of experience and therefore
a synthetic *a priori* necessity that phenomena should be intuited under
the form of time. (iv) The only way in which we can arrange phenomena in an objective temporal order is by conceiving them to be
entirely subject to causal laws. (v) Therefore the law of universal
causation is itself necessary as a presupposition of the possibility of
experience.

I think that there is a great deal in this argument. In particular, the
occurrence of series of impressions, which are reversible in the way that
Kant describes, is an essential factor in the construction of our picture of
the external world. As I have tried to show elsewhere, it alone permits
the common-sense conception of physical objects to be represented as a
theory with respect to a primary system of sense-qualia.[1] Nevertheless,
whatever other merits it may have, the argument does not achieve the
end for which Kant designed it; it does nothing to answer Hume. In
attacking Hume's problem, Kant began by making the same mistake.
He too assumed that if a principle of the uniformity of nature could be
demonstrated, it would serve to justify inductive inference; where he
differed from Hume was in believing that such a principle could be
demonstrated. But the truth is that such a principle can be demonstrated,
since it can be construed as analytic, and that it then serves no purpose

[1] See *The Origins of Pragmatism*, part iv, section 3C.

at all. For, as we have seen, it achieves its security at the price of lack of content. It is made consistent with anything whatsoever.

I ought to say at this point that I think that some content can be given to a principle of the uniformity of nature if it is taken as stating not that every event is subject to law, in the trivial sense that it can be brought under some true generalisation, but that every event is subject to what I call a manageable law, this being a true generalisation which is not merely supplied *ex post facto*, but one which is simple and fruitful enough for us to be able to use it in inferring the existence of unknown events. This is in line with Peirce's requirement that we should admit only hypotheses which relate to what he called *predesignated* properties. This concession, however, makes no difference to the present argument. In its revised form, the principle is too strong to be demonstrable; and it is still too weak to serve as the major premiss in Mill's syllogism.

The difficulty is that from a general principle of the uniformity of nature, or a general law of universal causation, even if it is so interpreted as not to be entirely vacuous, no specific laws are derivable, and Kant needs to be able to derive specific laws, or else his argument proves too much. It would, indeed, follow trivially from the assumption of determinism that the total state of the world at one time was determined by its state at another time, but to assume, at the level of Kant's examples, that any arbitrarily chosen earlier event was causally linked to any later event would entirely frustrate the search for causal laws. To obtain successful explanations at this level, we need to be able to establish particular causal chains; and this means that we must be able to treat the occurrences of certain sorts of temporally separated events as causally irrelevant to one another. Neither do we want to be put in the position of maintaining that we have to know all the laws of nature, before we can order events in time, for we do order events in time and we do not know all the laws of nature.

The truth seems to be that we do rely on *some* laws for the objective temporal ordering of events. Roughly speaking, the position which an event is assigned in an objective time order depends on the assessment of the position which it would appear to have to actual and hypothetical observers in different places, and in making these assessments we do rely on a number of laws – such laws, for example, as those concerning the velocity of sound and light. This does not mean, however, that the validity of these laws is guaranteed. They are open to falsification, and if in fact they were falsified, we should not necessarily be at a loss for a means of arranging events in an objective temporal order. The reason for this is that we are able, as it were, to hop from one foot to another. Our dating system of days and years, for example, depends on the uniformity of the earth's revolution on its axis, and of its revolution round the sun, that is, on the fact that the repetitions of these processes always occupy equal periods of time. Indeed, this sets our standard of temporal equality. Other processes take the same time, if they occupy the same quantity, the same fraction or multiple, of days or years. But this standard is not sacrosanct. It could be abandoned if it was found to lead to unsatisfactory results, that is, if it led us to regard processes as having equal duration, which we had other and better reasons for regarding as unequal. Then we should hop on to another foot, take advantage, let us say, of the law that the velocity of light is constant, and declare on this basis, or on the basis of some other periodicity, that the period of the earth's revolution was not uniform, or at least not uniform in the sense required. Our standards are always cross-checked and this means that no single one has to be regarded as infallible. A more simple example is that of the metre-rod. So long as a particular object, the metre-rod in Paris, was taken as a standard, it could be maintained that being a metre long just meant being the same length as this particular object. But it in no way followed from this that it was a necessary proposition that this object maintained a constant length. We could always admit the possibility that it might expand or contract. If there were physical reasons for supposing that this had happened to it, we should merely dethrone it from its position as a standard. Objects, the constancy of whose length had originally been established by reference to it, would now provide the measure by which its constancy would be tested and found wanting. This sort of circularity occurs all the time in science and is not in the least vicious.

Thus, no laws are sacrosanct, none is safe from rejection in the light of further experience, because, while we have to rely on some laws in building up our picture of the world, they do not always have to be the same ones. To use Neurath's overworked, but still valid, simile, there has to be a ship if we are going to remain at sea, but there is no part of the ship that we may not find it necessary to make over. At a given time, we cannot make over the part that is serving us as a foothold, but we can always move to another foothold and then make the previous one over. So there is no scientific hypothesis, no factual generalisation of any kind, and no presupposition, of which we can say that it is unassailable.

There is no guarantee even that we shall always find it possible to arrange events in a workable time order. It is indeed analytic that perceived simultaneities and successions can be correlated in some fashion or other, if we allow correlations of any degree of complexity, but by no means analytic that the correlations are simple enough for us actually to devise an objective system of dating. Our past experience may lead us to feel confident that our present system is not going to break down to any greater extent than is already provided for in the Theory of Relativity, but here again there is the inductive jump. Thus, it seems to me that Kant, for all the subtlety and penetration of his argument, provides no answer at all to Hume.

With the exception of John Stuart Mill, I believe that no philosopher of the first rank, since the time of Kant, has held the view that factual reasoning could be turned into a kind of valid syllogism with the principle of the uniformity of nature as a major premiss. On the other hand, it has been fairly widely held that if we are allowed to make suitable assumptions about the constitution of nature, we can prove that certain generalisations have a high degree of probability. Some of the philosophers who have taken this view have thought that they could escape Hume's charge of circularity; others have admitted it but held that it did not matter. But before we can go into this question, we must first try to clarify the difficult notion, or notions, of probability.

2 *A priori* Probability and the Frequency Theory

Before discussing any of the attempts which have been made to justify induction on the basis of the theory of probability, I have some points to make on the subject of probability itself. I take no side on the question whether there is one or more than one concept that goes under the name of 'probability', because I am not sure what the criteria are for individuating concepts and suppose that they are rather arbitrary. I do, however, believe that what we normally characterise as judgements of probability can be divided into three distinct classes, and if anyone wishes to infer from this that the word 'probable' is used in at least three different senses, I shall not quarrel with him.

The first of my three classes of judgements are those that relate to the calculus of chances. These may also be entitled judgements of *a priori* probability, though this appellation could be misleading, since there are philosophers who think, in my view wrongly, that judgements of my third class are *a priori*. Examples of judgements of my first kind are usually drawn from gambling. Thus, such judgements as that the probability of throwing heads three times in succession with a true coin is 1/8, or that the probability of throwing double six with a true pair of dice is 1/36, belong to this class.

My second class consists of what we may call statistical judgements, to the effect that such and such a character is in fact distributed in a given class with such and such a frequency. A typical example would be the judgement that it is probable to a degree m/n that such and such a recessive characteristic will appear in the *l*th generation of a given biological species, or the judgement that the probability that a man will die of lung cancer is increased if he is a heavy smoker.

Thirdly, there are what I propose to call judgements of credibility. These are judgements to the effect that some particular event is likely to happen, or to have happened, or that one event is more likely to happen than another, though the degree of probability cannot, or cannot in any obvious way, be given mathematical expression. Examples falling into this class would be the judgements that it is more probable that England will join the Common Market within the next three years than that Germany will be reunited, that it is more probable than not that Sir Philip Francis was the author of the Junius letters, that it is very improbable that such and such a picture is a genuine Leonardo, that it is not improbable that atomic tests can affect the weather. Judgements of credibility may be based upon statistical judgements, but they are not to be identified with them. Thus, the judgement which assigns to members of the class of cigarette-smokers such and such a probability of dying of lung cancer is, as I have said, statistical; but the judgement that such and such an individual smoker will probably die of lung cancer, if it is genuinely a judgement about this individual, and not just about the class of smokers to which he belongs, is a judgement of credibility. I shall have more to say about this distinction later on.

Care is also needed to distinguish judgements of my first class from statistical judgements. In the examples which I gave, the distinction depends on the way in which one interprets such expressions as 'a true coin' or 'a true die'. Thus, one might define a true coin as being one that satisfied certain physical conditions, with respect to its structure, or the location of its centre of gravity, or the material out of which it was made, and in that case our judgements about the probability of its coming up heads or tails would be statistical. They would be judgements about the actual relative frequencies of heads and tails in series of tosses which were made with coins that conformed to those physical specifications. Exactly how such judgements are to be analysed is another question into which I shall enter later on.

In the normal way, however, such examples are not interpreted in this fashion. A true coin, or a true die, is simply defined as one that yields results which are in accordance with the *a priori* calculus of chances. But then all these judgements become expressions of mathematical truisms. On the understanding that the coin has two faces, to say that the probability of throwing heads with a true coin is $1/2$ is just a way of saying that 1 is the half of 2. To say that the probability of throwing heads three times in succession is $1/8$ is to say that among all the possible ordered triplets of the numbers 1 and 2, such as 121, 211, 212 and so forth, the sequence 111 stands to the total in the ratio of 1 to 8. If we generalise this and say that the probability of throwing heads n times in succession is 2^n, what we are saying is that among all the possible ordered n-tuplets of the numbers 1 and 2, the sequence consisting of n 1s stands to the total number of possible sequences in the ratio 1 to 2^n. Evidently the value of $\frac{1}{2^n}$ diminishes as n increases, and on the assumption that roulette wheels run truly, this is all that there is to saying that a long run of either colour at roulette is highly improbable. The well-known Monte Carlo fallacy consists in assuming that since the odds against a red number's coming up n times in succession are very high, if n is at all large, the odds against red in the nth instance, when it has already come up n-1 times in succession, must also be very high. But this is an invalid inference. The calculation of the odds is based on the assumption that the successive spins of the wheel are mutually independent, so that the probabilities are the same in each instance, no matter what the results of its predecessors have been. For the sake of simplicity, let us assume that the game is fair and not, as it is in practice, weighted against the gambler by the presence of zero which counts against both red and black. Then even if red had come up a million times in succession, the odds against which are astronomical, the odds against red on the million-and-first turn would still be 1/2. This appears counter-intuitive until one realises that none of this talk about odds has anything to do with judgements of credibility, that is judgements about what is actually likely to happen, but is simply talk about abstract possibilities. To say that the odds against red's coming up a million times in succession are astronomical is to say that if one were to list all the possible million-term sequences of red and black, the sequence consisting of a million reds would be one out of an astronomically large number of alternatives. To say that the odds against red on the million-and-first turn are 1/2 is just to say once again that 1 is the half of 2.

The point here is that the calculus of chances is a branch of pure mathematics. We can use it to infer that if certain ratios obtain among a given set of possibilities, then, necessarily, certain other ratios obtain among them also. The references to pennies, dies, roulette wheels and so on are quite adventitious. What enables us to say, for example, that the chances of throwing double six with two dice are thirty-five to one against is the purely mathematical fact that there are just thirty-six combinations obtainable from the numbers 1 to 6 by pairing each number with itself and with one of the other numbers, and that the combination of 6 and 6 is one out of this total. The empirical information that a die has six faces and that dice are commonly so constructed that each face comes up about as often as any other in a long series of throws is quite independent of the validity of this sum, and conversely the fact of there being such and such possible distributions among the pairs of numbers in question – for example, the fact that there are five possibilities which yield a total point of 8, so that the chances of making a score of 8 with two dice are just worse than six to one against – tells us nothing about what would happen in any actual game of dice. If we are going to apply the calculus of chances to actual games, we have to make the assumption that all the logically equal possibilities are equal in fact, and this of course is not a mathematical truth. It is an assumption to which one has to give an empirical meaning.

Unfortunately, this is often overlooked. It is a common mistake for those who are writing on this subject to start with a well-defined notion of *a priori* probability, namely that of the incidence of certain features in the total of all possible mathematical combinations of a given sort, and then go on to talk of its being or not being equally probable that this or that possibility will be realised in nature, thereby using the word 'probable' in a way that their definition no longer covers. Mathematically, the alternatives 1-6, in the case of the dice, are equally probable in the entirely trivial sense that in the series 1-6 each of them occurs just once. But clearly this triviality cannot be applied to any actual game, unless one makes suitable assumptions about actual frequencies, and then introduces a new notion of probability, or at least gives the old notion a new application which will need to be defined.

B. KEYNES'S USE OF THE MULTIPLICATION AXIOM

A good illustration of the way in which this seemingly obvious point can be overlooked is to be found in the use which Maynard Keynes makes of the multiplication axiom of the calculus of chances, in his *Treatise on Probability*. Keynes is one of those who believe, in my opinion wrongly, as I shall argue later, that to say that something is probable is always to say that it is probable relatively to something else. He takes the terms of this relation to be propositions and symbolises it by means of the expression $P(a/h)$ which is to be read as 'the probability of a, given h'. However, as it has since become customary to use the letter h not for the evidence but rather for the hypothesis on which the evidence is supposed to confer probability, I shall, in reproducing Keynes's formula, substitute for it the letter e. Though it is, for most purposes, convenient to treat 'a' and 'e' in this formula as standing for propositions, we could also take them to stand for events or for classes. If we interpret the word 'probability' in the way in which it is applied to the calculus of chances, then the formula $P(a/e)$ may be understood to refer to the proportional numerical incidence of the case a in the domain of possible cases e. If a and e are taken to be classes, $P(a/e)$ becomes the proportion of members of e which are also members of a.

I come now to the multiplication axiom, which figures in all the standard axiomatisations of the calculus. In our symbolism, it runs: '$P(ab/e) = P(a/e) \times P(b/ae) = P(b/e) \times P(a/be)$.' If we construe this formula as stating that the proportion of the set of possibilities e which are combinations of a and b is equal both to the proportion of e which are a multiplied by the proportion of the combination of a and e which are b, and to the proportion of e which are b, multiplied by the proportion of the combination of b and e which are a, it is easy to verify that it is a mathematical truism.

Now Keynes's idea was that this axiom can be used to prove that the probability of a universal hypothesis, on given evidence, is increased by the addition to the evidence of any instance which is favourable to the hypothesis, in the sense that it satisfies both its antecedent and its consequent. Thus, if the hypothesis were of the form 'For all x, if fx then gx', every case which came up of something which was both f and g would make it more probable. This conclusion is not startling in itself: it is, indeed, a common assumption that a generalisation is confirmed by the discovery of instances in which it is exemplified. What is startling is that this conclusion should be held to follow from the axiom of multiplication. In fact, we shall find that there is nothing wrong with Keynes's proof, which is both simple and elegant. The question is only whether it proves what he thinks it does.

In setting out the proof, I shall follow the exposition which Jean Nicod gives of it in his book *The Foundations of Geometry and Induction*. Like Nicod, I shall adopt the usual practice of employing the letters p and q as propositional variables, in place of Keynes's a and b.

So, let p be a universal proposition, let q be a singular proposition exemplifying p, and let e be the existent evidence, which includes the fact that q satisfies the antecedent of p but not that it satisfies its consequent. And let it be assumed that the values of both $P(p/e)$ and $P(q/e)$ are greater than 0 and less than 1.

Then, by the axiom of multiplication, we have:

$$P(pq/e) = P(p/e \times q/pe) = P(q/e) \times P(p/qe).$$

Consequently,

$$\frac{P(p/e)}{P(p/qe)} = \frac{P(q/e)}{P(q/pe)}.$$

But since if a generalisation is true, it is true in every instance, and since the evidence includes the fact that q satisfies p's antecedent, p and e jointly entail q. Consequently, $P(q/pe) = 1$. But then, since *ex hypothesi* the value of $P(q/e)$ is less than 1, the value of $P(p/e)$ must be less than the value of $P(p/qe)$. In other words, the addition to the existing evidence of the fact that the generalisation has been verified in a new instance increases its probability.

Now there is nothing wrong with this, mathematically. On the assumptions that Keynes and Nicod make, which seem acceptable, their conclusion does follow from the multiplication axiom. It must, therefore, be open to an equally platitudinous interpretation; and it is easy to show that this is so. Let us, for the sake of simplicity, allow our generalisation to range over just three tosses of a coin and let the hypothesis be that all three tosses come up heads. Before the experiment, then, the probability of this hypothesis is 1/8, the probability of getting heads on the first toss is 1/2, and the probability of getting heads on the first toss given that the hypothesis is true is 1. So, all the requisite conditions are satisfied. Suppose now that the first toss comes up heads. What follows? We have a new quantity to calculate, namely the probability that all three tosses come up heads, given that the first one does. Now plainly the odds have shortened, since only four possibilities remain out of the eight with which we started, so that the verification of the hypothesis in one instance has increased its probability from 1/8 to 1/4. If heads comes up on the second toss, the same process is repeated. Only two possibilities remain, so that the probability of the hypothesis rises to 1/2. Finally, if heads comes up on the third toss, there are no adverse possibilities left; the probability of the hypothesis has risen to 1.

But this really is platitudinous. It is, indeed, necessary that the number of possibilities be finite, in order that the successive verifications should diminish it. But, provided that it is finite, the increase in the probability of the hypothesis simply consists in the fact that as the number of possibilities grows less, the proportion of the possibilities which are adverse to any given distribution of values is bound to lessen also. And this holds no matter what the probability of the distribution may be in any given instance. For it is true of any fraction x/y that the more times you multiply it by itself the smaller it becomes. Consequently, if the situation is changed by its having to be multiplied one less time, it inevitably grows larger. What all this comes to, in short, is the mathematical truism that when x is less than y and n is positive, $(x/y)^{n-1}$ must be greater than $(x/y)^n$. On the other hand, the value of x/y of course remains constant. The probability of the original hypothesis increases with each verification, but the probability at each stage of its being verified in the next instance remains exactly the same.

An important consequence of this last point is that in a system of this kind the probability of the general hypothesis, so long as its truth or falsehood is not decided, can never increase beyond the probability of its being verified in any given instance. For, as we saw when exposing the Monte Carlo fallacy, the length of the favourable run is immaterial. If there is only one case undecided, the *a priori* probability that the hypothesis is true is just the probability of its succeeding in this instance, and that, in a case where there are only two possibilities, is 1/2. It is, therefore, a mistake to talk, in the way that Keynes and Nicod do, as if the hypothesis could be made to approach certainty through successive verifications. At no stage of its journey does it ever come any nearer certainty than the probability of its verification in the very first instance. If it comes unviolated to the last hurdle, then either its probability jumps to unity – the race is over and it has won – or it falls to zero; the race is over and it has lost.

C. CARNAP'S SYSTEM OF INDUCTIVE LOGIC

It should, however, be said that this particular objection to Keynes's use of the multiplication axiom does not hold against every attempt to found induction on the calculus of chances. It is possible to arrange that the chances mature, or in other words that the accumulated credit of past verifications is passed on to the next instance. A good example of this is to be found in the system of inductive logic which has been developed by Rudolf Carnap, whose concept of what he calls logical probability is identical with Keynes's. I shall not attempt to give a detailed account of Carnap's rather intricate system, but its main ideas are simple and a summary of them will be sufficient for the criticisms that I wish to make.

One of the problems in applying the calculus of chances to empirical events is the assignment of initial probabilities. Within the calculus, as we have seen, the probability of an event is obtained by listing a finite number of equal possibilities and calculating the proportion of them which are favourable to it. In the examples which we have so far considered, the listing of the possibilities presents no problem, since it is achieved by a simple mathematical operation. But once we go outside the domain of pure mathematics, how are we to decide which events are equally possible?

There are two current ways of dealing with this problem. The first of them, which is employed in the application of the calculus to actual games of chance, is to count certain types of event as equally possible when we have reason to believe that they occur in a sufficiently long run with approximately equal frequency. This may not be an objectionable procedure in itself, but it is plainly a renunciation of any attempt to assess the probability of actual events on the sole basis of the calculus of chances. We move into the field of statistical judgements, with which I shall be dealing later on.

The second method of applying the notion of equal possibility to actual events is to decide the question semantically. This method has to be employed with care, since it can easily lead to contradiction. For instance, it has been held by some writers that in a case where we have no evidence either for or against a given proposition, we are entitled to assume that it is equally likely to be true or false. Suppose, then, that I am playing a game of drawing marbles from a bag, and that, relying on this principle, I take it to be an even chance that the first one to be drawn will be blue. This would be a foolish assumption to bet on but it would not be contradictory, so long as I treat not-blue as a single colour on a level with blue. But if I follow the natural course of breaking down not-blue into a disjunction of other colours, and if, by parity of reasoning, I also take it to be an even chance that the first marble to be drawn will be red, an even chance that it will be green, and so forth, then I do fall into contradiction. If there are more than two options, it is not mathematically possible that each of them should have an even chance of being realised.

The moral of this is that the predicates which are to furnish a set of equal possibilities must satisfy the following two conditions: they must, of course, be finite in number, for otherwise the proportion of favourable possibilities in the total could not be calculated, and they must be co-ordinate, in the sense that if f and g are two such predicates and if g

can be represented as a disjunct of two other predicates *h* and *i*, the disjunction of *f* with either of these predicates must not be taken as exhibiting equal possibilities. This seems a reasonable requirement in the example that I gave, but it can appear arbitrary, as is shown by a simple example which was devised by Dr J. L. Watling in an article entitled 'Confirmation Discomforted'.[1] He asks to suppose that 'we are following a man along a road and reach a place where the road divides into three, two paths climbing the hillside, one lying in the valley'. Knowing nothing but that the man, now out of sight, will take one of the three paths, how are we to estimate the probability that he will take the path lying in the valley? If we regard all three paths as equally possible, we shall estimate it at 1/3, but we might just as well regard his going into the hills and his going into the valley as being equally possible, in which case it would be 1/2. This does not show the method to be inconsistent, since we can always take measures to ensure consistency. In this instance, we have antecedently to make a ruling which will direct which procedure we are to follow. Even so, it appears unsatisfactory that different methods of ensuring consistency should lead to markedly different results, and that the decision between them should be wholly arbitrary.

This may, however, not be an entirely fair example, since it calls for an *ad hoc* decision rather than one which is in accordance with some systematic procedure. If we are to approach the question systematically we must have a universe of discourse in which we start with a number of predicates which are declared to be co-ordinate; presumably they will be chosen in such a way as to make this seem a reasonable decision: for instance, they might all appear to be equally specific. The assignment of these predicates in all logically possible ways will then determine a set of equal possibilities. Although it is not necessary, as we shall see, that these equal possibilities be regarded as being equally probable, still, however the probability is distributed, the result will be that its assignment to a given event will depend not only on what happens in the world, but also on the structure of the language which we are using to describe it. This may appear counter-intuitive but, so far as I can see, it is an inevitable feature of this type of approach.

This is anyhow the line that Carnap follows. His procedure is to construct a simplified model which it is hoped will prove capable of

[1] Published in the *Revue Internationale de Philosophie* (1963). I have also quoted this example in my essay on 'Chance' which was first published in the *Scientific American*, CCXIII, 4 (1965), and reprinted in my *Metaphysics and Common Sense*.

further elaboration, to the point where it can be applied to the actual world. For the materials of his model he takes an artificial language with an infinite number of individual constants, a finite number of primitive predicates, and the usual truth-functional connectives and quantifiers. In the earliest version of the theory, as it is set out in Carnap's book *The Logical Concept of Probability*, it was stipulated that the primitive predicates should be logically independent of one another, which was a very serious restriction. Not only did it exclude quantitative predicates but it meant that there could be only one predicate which fell under any one determinable: for instance, only one colour predicate or only one predicate of shape. I should, however, add that this restriction has subsequently been removed. In its later development, the theory allows families of predicates to be taken as primitive. But since my criticism bears on matters of principle rather than matters of detail, I shall summarise the theory in its earlier and more simple form.

On Carnap's original assumptions, it is evident that we can get a complete description of the universe corresponding to Carnap's universe of discourse, by stating with respect to each individual in it whether or not it has each of the primitive predicates. If the number of the individuals is infinite the description cannot be written out, but with respect to any finite sub-section of the universe it can be, though the process might be very laborious. Let us take as an example a very small universe with three individuals, *a*, *b*, *c*, and two predicates, say the predicates 'red' and 'round'. Then the statements '*a* is red and not round', '*b* is red and round', '*c* is not red and not round', describe a complete state of this universe. The statements '*a* is red and not round', '*b* is not red and round', '*c* is red and not round' describe an alternative state. Carnap calls any such assignment of the primitive predicates or their negations to the individuals of his system a state-description of the system.

Two such state-descriptions may differ, not in the distribution of their primitive properties but only with respect to the individuals to which these properties are ascribed. Thus, in our simplified example, the state-descriptions '*a* is round and not red, *b* is round and red, *c* is not round and red' and '*a* is not round and red, *b* is round and not red, *c* is round and red' satisfy this condition. When two or more state-descriptions stand in this relation to one another they are said to be isomorphic. A disjunction of state-descriptions which are mutually isomorphic is called a structural-description.

Carnap lays down the convention that in the case of any language representing a universe of discourse in which there is a finite number of individuals, every state-description receives from purely tautological evidence a degree of confirmation greater than 0. Since tautologies are construed as saying nothing about any matter of fact, this means that by convention every state-description is assigned some initial degree of confirmation. Furthermore, since every sentence in such a language can be written out as a disjunction of state-descriptions, it follows that an initial degree of confirmation is assigned to every statement which the language has the resources to express. The initial degree of confirmation of any molecular sentence is the sum of the initial degrees of confirmation of the state-descriptions which it disjoins. In the case of a logically true statement which disjoins all the state-descriptions this sum reaches a maximum of 1.

Given these conventions, it can be proved by the multiplication axiom that in the case of any language L_n, where the number n of individual constants is finite, the degree of confirmation c of a hypothesis h on evidence e is always equal to the initial confirmation of h and e divided by the initial confirmation of e. In the case where the number of individual constants is infinite, the degree of confirmation is defined as the limit to which the values of $c(h,e)$ converge, if they do converge, as n increases.

It follows that all that we need to do in order to be able to calculate the degree of confirmation which any given body of evidence lends to any given hypothesis is to assign initial degrees of confirmation to the state-descriptions. As Carnap sees it, this is purely a matter for decision. Our aim is to arrive at assessments of probability which are intuitively acceptable.

His own choice, in *The Logical Concept of Probability*, is to assign equal initial measures not to state-descriptions but to structural descriptions. From this he derives, as the initial degree of confirmation of any state-description S, the formula $\dfrac{1}{tz}$ where t is the number of structural descriptions in the language and z the number of state-descriptions which are isomorphic with S. This results in a system which is heavily weighted on the side of uniformity. The likelihood that a given individual possesses a given combination of properties is very much increased by the evidence that other individuals possess it. In this way, when a universal hypothesis has been found to be satisfied in a large number of instances, a high degree of confirmation accrues to the proposition that the next instance will also satisfy it. On the other hand, it is a somewhat startling feature of Carnap's system that the probability that a universal hypothesis holds in a universe with an infinite number of individuals, no matter what the evidence, is always o. This has the strange consequence that a kind of ontological argument is valid in the system. If we take any non-contradictory predicate which it contains, whether simple or compound, then, provided that the number of individuals is infinite, we can demonstrate that the universal hypothesis that all of them lack the property for which the predicate stands has the probability o; and from this it will follow that the existential hypothesis that at least one of them has this property has the probability 1.

This conclusion is not at all easy to accept if judgements of probability, in Carnap's sense, are to be identified with judgements of credibility, though I shall try to show later on that it is a not very startling offshoot of the idea of confirmation with which he is actually operating. I do not, however, wish to criticise Carnap on points of detail. For one thing, he continued to emend the system until the end of his life, and his followers are still emending it, so that any objection to a point of detail runs the risk of being out of date. And, in any case, my concern is not with the merits or demerits of Carnap's assessment of the initial degree of confirmation which he chooses to assign to his state-descriptions, but with the very notion of an initial degree of confirmation itself. What can possibly be meant by speaking of a statement as having any degree of confirmation relatively to a tautology, or in other words, antecedently to any evidence?

The answer is that when Carnap speaks in this way he is making a purely formal assertion. In terms of the measure function which he chooses, to say that a statement has such and such an initial degree of confirmation is simply to say that it disjoins such and such a number of state-descriptions and that these state-descriptions themselves disjoin such and such a proportion of the total number of structural descriptions. But what are we then to make of the assumption that every structural description has an equal measure, if this is to be understood not just as the truism that in the enumeration of structural descriptions each one counts as one, but as a guide to our expectations of what will actually happen? At this point, the purely formal concept of probability with which Carnap has been working needs to be given a material interpretation.

The nearest Carnap comes to fulfilling this need is in his talk of fair betting quotients. Roughly speaking, the probability of a hypothesis *h* on evidence *e* is equated with the odds at which it would be rational for one to bet on *h*, in the light of one's knowledge of *e*, where *e* is one's total stock of relevant information. Later on, I shall show that there is a strong objection to regarding anything of this kind as a definition of probability, in the application of the term to credibility judgements, but at present I am concerned only with the problem of attributing some empirical content to Carnap's choice of a measure-function. So now we must ask what determines the rationality of a bet. The answer which Carnap gives to this question is, in effect, that a bet is rational when it is at least actually fair, whether or not we have good reason to think it so, and that a bet is actually fair when the odds match the actual frequency with which the kind of event on which one is betting occurs in the kind of situation which is covered by the evidence.

This suggests that we should interpret the assignment of equal initial degrees of confirmation as a postulation of equal actual frequencies, but unfortunately such an interpretation would not be applicable to the unchanging universe to which Carnap's language relates. The facts in this universe are determined once for all by the truth and falsehood of the various state-descriptions; they uniquely determine its structure, so that any talk of frequency with respect to the incidence of structural descriptions will be devoid of sense. Of course, when I speak of this universe as unchanging, I do not mean that it cannot be in different states at different times. The point is just that this is already provided for in the catalogue of state-descriptions – in the most simple way, by taking the individuals to be spatio-temporal.

We must look elsewhere, then, for an empirical interpretation of the assignment of equal measures to structural descriptions, and we can find it in the assumption, which underlies this assignment, that the more state-descriptions are isomorphic to a given state-description, the smaller its measure has to be. This has the effect of strongly favouring an unequal distribution of the primitive properties, since it postulates that they are either widely present or widely absent. For instance, in our example of a very simple universe with just three individuals and two primitive properties, the state-descriptions in which all the individuals have both, or neither, or just the same one of the primitive properties are, if my calculations are correct, awarded six times the initial degree of confirmation that is awarded to the state-descriptions in which no two of the individuals have the same combination of positive and negative properties. Not only that, but this bias is very strongly projected. In the original version of Carnap's theory, the probability that an unexamined instance has a primitive property P is given by the formula $\dfrac{S' + W}{S + K}$, where S is the number of individuals so far examined, S' is the number of those which have been found to have the property P, K is the total number of predicates, called by Carnap q-predicates, which are obtainable by conjoining the primitive predicates and their negations, and W is the number of q-predicates which the predicate designating P disjoins. The influence of the factors W and K will depend upon the richness of the language, but, however rich the language, it will tend to become negligible as S increases to infinity. Consequently, so long as all, or even very nearly all, the examined instances remain favourable, the probability that an unexamined instance will be favourable approaches unity. And what this must be taken to mean, when it is translated into betting terms, is that if we take a set of primitive predicates, each of which has been exemplified by S' out of a total of S examined individuals, and wager with regard to each of the predicates that a hitherto unexamined individual will exemplify it, we shall be right in the proportion yielded by the formula – with the result that, under the conditions which I have set out, we may come to the point where we practically always win.

But now we can see that what Carnap has done is simply to build into his system a very strong assumption of the Uniformity of Nature, an assumption which he does not attempt, and indeed has not the means to justify. It is not an empty assumption, first because the range of the uniformity is circumscribed by the choice of predicate primitives, and secondly because the incidence of the uniformity is given a numerical measure. It is, however, an insensitive assumption, since any pervasive feature of a considerable segment of space-time is projected with equal force. This process is, indeed, self-correcting, in the sense that the odds are adjusted if the bets go wrong. But all that this means is that failures are projected as mechanically as successes. The decision as to what to project is already taken with the construction of the language, and considerations of the quality, as opposed to the quantity of the evidence play no part at all.

D. THE LAW OF LARGE NUMBERS

I shall not try to give an account in any detail of the various other attempts that have been made to extract from the *a priori* calculus of chances conclusions about what is actually likely to happen. Not surprisingly, it is a feature of all of them that at some stage or other an empirical assumption is tacitly introduced. Very often, this is done under cover of the Law of Large Numbers. The upshot of this law, which is mathematically demonstrable, is that in the case of any sufficiently large sample which is drawn from a larger population, there is a high probability that the ratio in which a given character is distributed in the sample approximately matches its distribution in the parent population: as the size of the sample increases, this probability approaches unity. Once again, however, one must be clear about what is meant by probability in this connection. Since the law of large numbers is a theorem in the calculus of chances, the reference to probability here can only be construed in terms of the incidence of logical possibilities. To say that if the sample is sufficiently large, there is a high probability that it approximately matches the parent population, with respect to the distribution of any given character, is just a way of saying that if we take all the possible selections from the parent population which yield a sample of its size, we shall find that those which roughly match the parent population in the respect in question very greatly outnumber those which do not. In short, deviant samples are untypical, and become more untypical the larger the sample.

We have, then, if we make our samples large enough, the mathematical assurance that the large majority of possible selections approximately match the total population from which they are drawn. So setting aside the case, which is of no interest to us, where we know that our sample constitutes so large a fraction of the population as to swamp the unexamined instances, we can reasonably infer from the composition of the sample to the composition of the total population, provided that we can reasonably assume that our sample belongs to the undeviant majority. And it is easy to show that we can reasonably assume this if we are entitled to make the assumption that any one sample of a given size is as likely to be selected as any other. This last assumption is, I think, invariably made, either explicitly or tacitly, by those who pursue this line of argument. When it is made explicitly, it is presented as an innocent truism. And it is indeed an innocent truism in the sense that if we equate the probability of a particular selection with the ratio which it bears to the total number of possible selections, each selection counts for one and only one. An omniscient being who made every possible selection would necessarily find that most of his samples were typical.

It hardly needs saying, however, that we are not in this position. The Briareus image in which the population is sorted into bags, and we simultaneously dip a hand into each one, is a very good model for the operation of fair sampling, but it does not fit the facts. So far from its being the case that we are as likely to make any one selection as any other, there is a vast number of selections, indeed in most instances the large majority, that it is impossible for us to make. Our samples are drawn from a tiny section of the universe during a very short period of time. And even this minute part of the whole four-dimensional continuum is not one that we can examine very thoroughly.

For these reasons, we need to make two quite strong empirical assumptions. They are first that the composition of our selections, the state of affairs which we observe and record, reflects the composition of all the selections which are available to us, that is to say, all the states of affairs which we could observe if we took enough trouble; and secondly that the distribution of properties in the spatio-temporal region which is accessible to us reflects their distribution in the continuum as a whole.

The first of these assumptions may be fairly readily conceded. It is, indeed, the purpose of scientific method to make sure that it holds. The reason why instances are not only multiplied but also varied, the importance of testing hypotheses under different conditions, is, so far as possible, to prevent our selections from being biased by the circumstances in which they are made.

The second assumption is the one that is troublesome. For clearly it postulates the uniformity of nature, not trivially but in quite a strong fashion. Indeed, it goes beyond what we naturally believe. We should not be very surprised to learn that many matters were very differently ordered in other parts of the universe, or even that the earth and the behaviour of the things upon it were going to change very radically in the next million years. For instance, it would not at all astonish us if any of our taxonomic generalisations were to break down in some remote region of space or time, however strong a local sample we had been able to collect.

But, it may be said, it does not much matter to us what goes on in the outer galaxies or what is going to happen on earth in a million years' time. We are seriously interested only in our fairly close neighbourhood and in the fairly immediate future. After all, the main purpose of inductive logic is to establish the reliability of hypotheses upon which we have to act. Consequently, the fact that our selections may be biased with respect to the universe as a whole is not of any great importance.

I think that this point is well taken, but it does not save the argument. For now we are faced with another difficulty. If we confine our attention to our familiar environment, and to the relatively near future, then, so long as the rate of growth of the populations in which we are interested does not substantially increase, we can be certain, and that without making any further assumptions, that in many cases the percentages with which the characters for which we are sampling will be distributed among them will not be very different at the end of the future period from what they are now. This will be true in all those cases in which we have built up such a backlog of instances that they are bound to swamp the new instances, however deviant these may be. But this conclusion is of no value to us. For we are interested in the maintenance of a percentage only in so far as it affects the new instances. We do not want to be assured that even if these instances are deviant the final result will be much the same. If we make the time short enough, we know this anyway. We want to be assured that the new instances will not be deviant. But for this we do require a non-trivial assumption of uniformity. Moreover, as we saw when discussing Carnap's theory, we are left with the problem not only of justifying any such assumption, but also of formulating it in such a way that it licenses the special inferences that we wish to make. We need to be shown what properties it is reasonable to project and in what circumstances it is reasonable to project them. But now it seems obvious that no general principle of uniformity could do this work, even if it could be justified. I shall return to this question at a later stage.

E. THE FREQUENCY THEORY

I turn now to my second class of probability judgements, consisting of those which I have characterised as statistical. It is to judgements of this kind that the frequency theory of probability is especially adapted.

The essence of the frequency theory is that it identifies the probability of an event with the proportion of instances in which the property by which the event is identified is in fact distributed among some class, of which the individual to which the property is ascribed is a member. Thus, if I say that the probability that some child which is not yet born will be a boy is 51 per cent, I am taken to be saying that in some class to which this child will belong, say the class of children born in England in the 1970s, the proportion of boys is 51 per cent. This statement contains the prediction that the child will be born, and in the event that this prediction fails, it is a matter for decision whether the statement of probability should be said to be false or inapplicable. If we choose to say that it is false, we shall be treating probability statements of this kind as conjunctions with an existential component. They will be taken to assert both that there is an event e of character C which belongs to some class K, and that the distribution of C throughout K is in such and such a proportion m/n. The statement can then be falsified in two ways, either by Ks turning out not to contain e, or by its exhibiting C in a different proportion from m/n. Alternatively, we can treat the truth of the existential component as a presupposition, and make the statement's truth or falsehood depend only on the truth-value of its probability component. This, however, has the rather odd consequence that what the statement is understood to assert, as distinct from what it pre-supposes, does not refer to the individual event at all, but only to some class to which the event belongs.

The probability component in my example also contains a prediction, inasmuch as the reference class, in this case the class of children born in England in the 1970s, includes future instances. This need not always be so. A demographer who is recording changes in vital statistics may, for example, speak of the probability that a child born in the eighteenth century would live to the age of forty, perhaps in order to contrast it with present-day life-expectation. In such a case his statement would not be predictive; it would simply be a minuting of an ascertained ratio, the proportion of children born in England in the eighteenth century who did attain the age of forty. It is, in a different sense, improbable that this figure would be more than very roughly accurate, but theoretically the record could be complete. On the other hand, the majority of statistical statements which there is any point in making are at least to some extent predictive.

An important point to notice is that the probability of an event, in this sense, may be different, according as it is assigned to different classes of reference. A child born in England in the 1970s is also a child born to parents who have such and such occupations, a child who is going to grow up in such and such an environment, and so forth. In relation to the child's sex, it may, indeed, be assumed that the choice of one or other of these reference classes does not greatly matter. Provided that they are large enough, the sex distribution in each of them is likely to be much the same. But this is by no means true universally. It does not, for instance, apply to life-expectation, where the heredity of the child, the prosperity of its parents, the environment it grows up in, make a big difference. I shall return to this point later on.

As I have so far stated it, the frequency theory applies only to finite classes, since it is only with respect to a finite class that one can speak, in any straightforward sense, of definite percentages. In a good many instances, this restriction does not matter. Clearly the class of children born in England in the 1970s is going to be finite, and it would appear safe to assume that this also applies to the class of all children born any-where at any time. On the other hand, there are cases in which we might not wish to be committed to this assumption. For example, many scientific laws now take a statistical form. There is said to be a probability of $1/2$ that a radium atom disintegrates within 1700 years, and it seems undesirable that those who put forward this hypothesis should be taken to be committing themselves to the proposition that the number of radium atoms is finite. There is also the point that a scientific law, even one of a statistical character, may be credited with a scope that goes beyond its actual instances. It is supposed to entail subjunctive conditionals and this is sometimes taken to mean that it also holds good in merely possible cases. Thus, the Mendelian laws of heredity may be thought to range not only over the organisms that there actually happen to be, but also over all possible organisms of which the number is presumably not finite. I am, indeed, doubtful, as I have already said, about the necessity or purpose of this assumption even in the case of causal laws, but I shall not pursue this question here.

For one or other of these reasons, modern exponents of the frequency theory have made it applicable to infinite classes. This is achieved by introducing the notion of a limit. It is held that to attribute a probability m/n to an event with the character C is to identify the event as belonging to some class the members of which can be arranged in a serial order, such that there is a point in the series after which the proportion of those that have the character C, counting from the beginning, does not differ from m/n by more than ϵ, where ϵ is as small a fraction as you please. Thus, if I say that the probability of a penny's coming up heads is $1/2$, where this is to be understood as an empirical statement about the behaviour of a particular coin and not as a deduction from the calculus of chances, what I am supposed to be asserting is that in a possibly infinite series of tosses of the penny there will come a point after which the total percentage of heads will not differ from $1/2$ by more than ϵ. And this is what is meant by saying that the series of tosses tends to the limit of 1/2 with respect to the incidence of heads. If the event in which we are interested belongs to no class which tends to a limit in this way, this notion of probability is not applicable to it.

In its best-known versions, the frequency theory is further complicated by the requirement that the series in question be randomly ordered. This is thought necessary in order to prevent artificial arrangements which would deprive a series of a limit which it would otherwise have attained or burden it with alternative limits. Thus, if the figure 1 is taken as representing heads and the figure 0 as representing tails, an infinite sequence of tosses conforming to the pattern 0101101110111, where the appearance of every 0 is followed by an increasing number of 1s, will tend to the limit of 100 per cent heads; a fact from which it follows, incidentally, that 100 per cent frequencies, in this sense, cannot be identified with causal laws, since a causal law admits of no exceptions, whereas a 100 per cent frequency, as this example shows, can accommodate an infinite number of them. On the other hand, if we arrange the sequence in such a way that the 0s and 1s alternate, which will clearly be possible since there is an infinite number of both, it will attain the limit not of 100 per cent but of 50 per cent heads; and if we put all the 0s first it will fail to attain either limit, unless we suppose ourselves capable of coming to the end of an infinite series.

These awkwardnesses are avoided if we confine ourselves to sequences which are random in the sense, defined by von Mises,[1] of being indifferent to place-selection. What this requires is that every subsequence which is obtained by such methods as selecting every third item, or every item corresponding to a prime number, or, in an example such as ours, every item immediately following a 1, or every item immediately succeeding two zeros, and so forth, should attain the same frequency limit as that attained in the main sequence. This requirement has been shown to be consistent, but it is so strong that it must always be doubtful whether it is satisfied. Professor Popper has, indeed, devised a method for constructing sequences which satisfy a similar but rather weaker condition of randomness,[2] but the doubt remains whether any empirical sequence meets his specifications, at least so long as it is not completed. I shall not pursue this question here, since the criticisms which I have to make of the frequency theory would still hold even if the requirement of randomness were not maintained.

[1] Cf. Richard von Mises, *Probability Statistics and Truth*, Lecture 1.
[2] See K. Popper, *The Logic of Scientific Discovery* (1934; English translation 1959) chap. viii, sections 51–9.

The first point to be noted is that when statements of probability are interpreted in this way, they are not falsifiable, so long as the series in question are not known to be complete. This follows from the fact that the failure of any such series to converge to a given limit at any given stage does not exclude the possibility of its doing so later on. However others may regard it, this is a serious problem for Professor Popper, who is committed to holding both that a statement must be falsifiable if it is to be accounted scientific, and that statements of probability, as they occur in science, are to be construed in accordance with the frequency theory. He handles this difficulty with rather too much *insouciance* by introducing a methodological rule which lays down the conditions under which probability statements are to be counted as being falsified. The rule is that they are to count as being falsified when they come up against reproducible effects in which the deviation from the postulated frequency is improbably great, in the sense defined by the law of large numbers. We shall, indeed, see that a rule of this kind has to be introduced if the frequency theory is to be made operative, but it is an awkward step for Popper to take, since the principle of falsifiability which he uses to demarcate science from metaphysics is couched in logical terms. For a statement to be falsifiable it has to be formally incompatible with some class of basic-statements; and this is a condition which probability-statements, as he construes them, do not satisfy. What he does, in order to keep them in the fold, is introduce a new criterion, while continuing to speak of falsifiability, as though the old criterion had not in this instance been abandoned.

While it would be generally admitted by exponents of the frequency theory that the judgements of probability to which this theory is adapted are not strictly falsifiable, they do regard them as being confirmable. If we come to a point where the recorded frequencies continue to cluster around a given value, we have support for the hypothesis that this is the limit which the series in question attains. Here again, however, it has to be admitted that the evidence is equally consistent with the hypothesis that the series eventually reaches quite a different limit, or even that it fails to reach any limit at all. In a series of tosses, a run of however many heads is consistent, as we have seen, with the limit's being 100 per cent tails, if the subsequent run of tails is sufficiently prolonged.

So far as the question of confirmation goes, this difficulty is not peculiar to the frequency theory. We shall see later on that on the received accounts of confirmation it is true in general that any evidence

which confirms a given hypothesis *h* equally confirms hypotheses which
are incompatible with *h*. Nevertheless, if we are looking to the frequency
theory to supply an adequate interpretation of statistical judgements, I
think that it is a serious objection to it that even if one knew what the
frequency-limit was with respect to the distribution of some character
in a given series, one could not deduce what the frequency would be
in any of its segments. This follows from the fact, which we have just
noted, that, if no term is set to the continuation of the series, the
eventual attainment of a given limit is consistent with the attribution to
any segment of any composition whatsoever. But it is chiefly in segments that we are interested. We want to project an ascertained
frequency over an ensuing set of instances. The best that we can do,
therefore, is fall back on the law of large numbers. In the normal case
in which we do not know what the frequency-limit in question is, or
even whether there is one, we can infer that if there is a limit which is
markedly different from that which we have recorded, the instances
which we have examined must be deviant; and we can then argue that it
is improbable that these instances are deviant and on the same grounds
that it is improbable that the ensuing instances will not exhibit much the
same frequency. But once more we have to find a basis for this judgement of probability, beyond the merely formal fact that deviant
instances are a minority of all possible instances. We have, it would
seem, to introduce some form of a fair sampling postulate. But if we are
going to rely upon a fair sampling postulate we do not need the notion
of frequency-limits. We can operate more simply with the model of
drawing balls from a bag.

This connects with the point, which I have already indicated, that in
the case of most of the statistical judgements that we actually work
with, the interpretation of probability in terms of frequency-limits is
too far-reaching. An insurance company is not concerned with the
accumulation or the projection of statistical data over any lengthy
period of time. In fixing its premiums, it takes no account of the vital
statistics of the seventeenth century or of what it conjectures that they
will be even fifty years from now. It is interested only in making rather
narrow extrapolations from contemporary data. And this applies
generally to the use of statistics in the study of human behaviour.

F. AN INTERPRETATION OF STATISTICAL JUDGEMENTS

Accordingly, I suggest that statements of probability, of the statistical kind, should be divided into two classes. The first class consists of those, like the Mendelian law of heredity, or the statistical laws of quantum physics, or the laws of the molecular theory of gases, that are built into some scientific theory. These are unrestricted. They state with respect to the distribution of certain characters or the behaviour of certain types of particles, that such and such ratios are regularly maintained, without setting any limits to the extent of the populations with which they are concerned. Following Professor Braithwaite,[1] I hold that statements of this kind are best interpreted according to the model of drawing balls from bags. They can be taken as implying that if we take samples of various given sizes, the frequency with which they exhibit the property in question will not be found to differ from the assigned ratio by more than such and such a relatively small amount. This has the corollary that if deviation from the frequency which a statement assigns is highly improbable, in terms of the law of large numbers, the statement will count as having been provisionally falsified. Its falsification will be only provisional, because we may go on to collect a large number of further samples in which the frequency differs only minimally from the one assigned; and in that case we can treat the single wild instance as being allowed for by the theory.

This means that we are building a fair sampling postulate into the interpretation of these probability statements. Our procedure is not circular because we are not using the postulate to guarantee either that the frequency which some character exhibits in a given sample matches its distribution in the parent population or that the same frequency will be exhibited in other samples of the same size. As Professor Braithwaite makes clear, we use the postulate simply as a rule of rejection. If we come upon deviations which are outside the limits set by the postulate, we conclude that the probability is not what our hypothesis has represented it as being. In certain cases, we may be able to adjust the hypothesis so that the deviations cease to be such.

If we ask how probable it is that such adjustments will be called for, then so long as we are interpreting probability-statements in this fashion, our question will have to be construed as a question arising out of some higher-level hypothesis about the frequency with which such adjust-

[1] See R. B. Braithwaite, *Scientific Explanation*, chaps v and vi.

ments have to be made. But since we can then raise a question about the probability of having to make adjustments to *this* hypothesis, and so *ad infinitum*, it becomes clear that this interpretation of probability-statements, though I believe it to be acceptable as an account of the way in which we do operate with a certain type of statistical generalisation, does not supply us with an analysis of statements of credibility.

So much for statistical generalisations of my first class. The second class consists of statements like 'There is a 51 per cent probability that his forthcoming child will be a boy' or 'The average contemporary Indian has a less than even chance of living to be forty', which are restricted in a way that statements belonging to my first class are not. These statements of my second class are difficult to analyse because they are hybrid and also in some respects vague. They are hybrid in that they both state a recorded percentage and predict its continuance and they are vague in that they fail to make clear what the scope of the recorded percentage is. When one speaks of the chances of the child's being a boy, one omits to specify the geographical and the temporal range of the statistics on which one is relying. Clearly, however, this is an omission that can easily be made good. A further source of vagueness, which is not removable except by a somewhat arbitrary ruling, lies in the failure to determine the range of the prediction, though here again it should be taken to be relatively narrow. Statistical statements of the kind here in question are comparatively unadventurous and by the same token strongly protected. If the sex ratio of births, say in Western Europe, were suddenly to change for some physiological reason which might or might not be known to us, so that the proportion of male children dropped to 30 per cent, we should not feel bound to say that the man who had estimated his chances of having a son at 51 per cent had been mistaken. We might regard him as having made a true statement about the statistics that prevailed at the time at which he was speaking. This is because such statements are construed as being predominantly summative and only marginally predictive. What is anomalous is that for the most part it is their predictive element which gives them their point. An insurance company is interested in past statistics only as a basis for estimating future frequences.

This account of statistical judgements exhibits a legitimate use of the term 'probability' but also one with important limitations. For instance, so long as we identify probability with relative frequency, we cannot significantly speak of probabilities as being affected by evidence, or of their increasing or decreasing. For if we are dealing with the incidence of some property in an infinite series of events, either the ratio in question does attain a limit or it does not. If it does, then this limit is the value of the probability all along, whatever our evidence may be at any stage; if it does not, then the notion of probability, in this sense, does not apply. If we are dealing with a finite class, the frequency with which a property is distributed among its members must have some fixed value, whether we are able to discover it or not. We do, indeed, say such things as that the probability of dying from tuberculosis in England is smaller than it was a hundred years ago, but this has to be construed as a comparison of two different probabilities, relating to the incidence of the same property in two different reference classes, say those of the population of England in the third quarter of the twentieth century and the third quarter of the nineteenth. Since these probabilities do not change, they are not affected by evidence. New evidence may cause us to change our hypothesis about the value of such a probability, but that is all.

A more serious limitation on the frequency interpretation of judgements of probability is that it construes them as referring only to classes and never to individuals, except in so far as they imply or presuppose that the individual in question is a member of some class. This is not always made sufficiently clear. A prospective father is told that there is a slightly better than even chance that his child will be a boy. Why? Because in some class to which the child will belong the ratio of male to female births is estimated at 51 per cent. But this is an identical inference. To say that there is a slightly better than even chance that this child will be a boy is just to repeat the assertion that the ratio of male births in some class to which the child is being assigned is just over 50 per cent. The only stage at which the individual comes into the picture at all is in his assignment to a class. From there on, he figures literally as a cipher. And there is worse to come. For an individual may belong to many different classes in which a given property occurs with different frequencies. To use a well-known example which was devised by Professor Cooley and quoted by Professor Hempel in his essay on 'Inductive Inconsistencies',[1] suppose that we ask what is the probability that Petersen, a Swede, is a Protestant. Well, 95 per cent of Swedes, let us say, are Protestants, so the odds are 19 to 1 on. But Petersen made a pilgrimage to Lourdes last year and 95 per cent, let us say, of those who make pilgrimages to Lourdes are Roman Catholics, so that the odds are at least 19 to 1 against. This is a flat contradiction, if we construe our original question as a question about Petersen's chances. It is not a contradiction, however, if we interpret the question in the only way in which it can be interpreted if we are identifying probability with relative frequency. For then our answer is just a restatement of the facts that Petersen is a Swede who has made a pilgrimage to Lourdes and that 95 per cent of Swedes are Protestants and 95 per cent of such pilgrims are Roman Catholics. There is no contradiction here; but equally the question which we were trying to ask about Petersen has disappeared.

But very often we do want to raise a question of this sort about an individual case. For instance, I may be interested in the statistics of longevity only because I want to know what *my* chances are of living to be an octogenarian. How then shall I proceed? What shall I take as my class of reference? The rational course would seem to be for me to choose among the many classes to which I belong the narrowest class, that is, the class defined by the largest combination of predicates, within which I think that the property in which I am interested occurs with an extrapolable frequency: so, in this case, not the class of all human beings throughout recorded history, or even that of contemporary Europeans or contemporary Englishmen, but rather that of contemporary Englishmen pursuing a certain type of occupation, having such and such a physical constitution, such and such dietetic habits, and so forth. As a rational man, I do not want to neglect any factor that is relevant to my case, and therefore will try to avoid choosing a reference class for which such a factor is not a defining property. There is, however, no place for this policy within the frequency theory. Indeed, the principle on which I am acting cannot in its terms even be formulated. One can attach no sense, within the limits of the theory, to saying that the choice of a narrower reference class yields a better estimate of my chances. For if to speak of my chances of having a certain property is merely an elliptical way of saying that the property is distributed in such and such a ratio throughout some class to which I belong, then, provided that the ratio has been correctly estimated in each case, any one statement of this kind is as good as any other. The choice of different reference classes gives different information, and that is all there is to it.

The upshot of this is that the judgements of probability of which the frequency theory gives an adequate account are those in which we are not concerned with individual cases. So long as the estimated proportion of its clients die in any given year, it does not matter to the insurance company who dies and who lives; so long as the estimated proportion of the descendants of such and such a set of organisms display a recessive characteristic in the nth generation, it does not matter to the biologist which members of a particular family have it and which do not, any more than it interests the physicist to know which electrons change their orbit, so long as the proportion is constant, or any more than the traffic manager cares whether it is the Smiths or the Joneses who travel on a Bank Holiday. This is not to say, of course, that we do not make use of ascertained frequencies when we estimate the probability of individual events, but only that probability judgements of this kind are not fully interpreted by the frequency theory. To interpret them fully, we have to relate them to probability judgements of my third class, that is, to those that I have been calling judgements of credibility. We shall find that they present an even more difficult problem of analysis.

3 The Problem of Confirmation

We have seen that judgements of probability, in the statistical sense, turn out to be elliptical. When, in their case, we talk of its being probable, to such and such a degree, that some individual has a given property, we are talking about the frequency with which the property is distributed among the members of some class to which the individual belongs. Probability of this sort is relative to evidence, in the sense that we might obtain a different, but no less correct, assessment of the probability, if we assigned the individual to a different class of reference.

Is it also true of judgements of probability of my third sort, those that I have called judgements of credibility, that they are relative to evidence? It has been widely assumed that it is. In particular, this is the view taken by those who regard the concept of probability, in this use of the term, as a logical concept, whether, like Keynes, they think it capable of accounting for every type of probability-judgement, or whether, like Carnap, they make separate provision for judgements of relative frequency. They take it for granted that when, in making a judgement of credibility, we assert that p is probable, part of what we are asserting is that p is made probable by some other proposition q, or some other set of propositions, q, r, s. What else, if anything, we are thought to be asserting is apt to be left unclear. It would seem that, on this view, we must at least be implying that the evidential propositions, q, r, s are true, perhaps also that we know them to be true, or at any rate that they are in their turn made probable by some other propositions which we know to be true. I suppose, however, that it could be held that the truth of the evidential propositions was presupposed, rather than implicitly asserted. In discussing this theory, I shall call the proposition that p is made probable by whatever evidence is in question the probability component of the proposition that p is probable, leaving open the question whether the theory requires that any other components be added to it.

Presumably, one reason why people come to hold a theory of this kind is that they see that to characterise a proposition as probable is not to assign it a truth-value somewhere between truth and falsehood. A proposition is either true or false; an event either occurs or it does not. There is nothing in between. But then, if an event is neither probable nor improbable in itself, it is tempting to infer that it must be probable or improbable in relation to other events. The consequence is, if probability is taken to be a logical relation, that all judgements of probability of this sort, or at any rate their probability-components, become analytic. For whether and to what extent one proposition probabilifies another will then depend primarily on the nature of the propositions in question, which is to say, the meaning of the sentences that express them, and apart from that only on the initial probabilities which may be assigned to them *a priori*. It is for this reason, indeed, that this theory is known as the logical theory of probability.

But now we come upon a difficulty which is analogous to that which arises when one tries to apply the frequency theory to an individual case. The analogy is between the choice of different classes of reference and the choice of different items of evidence. The logical probability of a proposition p, relative to q, may be different from its probability, relative to r, and different again from its probability, relative to q and r. Suppose that these different probabilities are known to us, and that we are concerned with placing our bets on the event described by p. Have we any reason, on the logical theory, to prefer one of these estimates to another? The answer is that we can have none. I am assuming that we have made a correct assessment of the strength of the relation in each case, so that all the competing judgements of probability, the judgement that the probability of p, given q, is m/n, that the probability of p, given r, is m'/n', and so forth, are equally true: indeed, if these relations are logical, they must all be necessarily true. But then how can we decide between them? How can one necessary proposition be better than another?

The answer which Carnap and his followers make to this objection is that we can adopt the methodological principle that the judgements of probability which we apply to particular events are to be those that are relative to total evidence. This is a plausible answer, but I shall show that it does not meet the difficulty. To begin with, it is not clear what the total evidence is meant to be. If we take it to be the totality of true propositions, then the probability of any hypothesis becomes either 0 or 1, since the totality of true propositions must include either the hypothesis itself or its negation. There are, indeed, ways of limiting the totality so as to avoid this conclusion – for instance, it could be required that the evidence be such that neither the hypothesis in question nor its negation followed from it – but they all encounter the further objection that they leave us with a methodological principle to which we should rarely be able to conform, since even under the restrictions which they imposed, there would usually be relevant evidence that was not in our possession. No doubt it is for this reason that Carnap goes to the other extreme. He appears to equate the total evidence with the totality of propositions which are in fact accepted by the person who is making the judgement. Perhaps his speaking of these propositions as evidence is meant to carry the implication that they are true, though here again it is hard to see how this requirement could be written into a methodological principle. The most that could be asked, it would seem, would be that one should relate one's judgements to the totality of propositions which one believes that there is good reason to think true; and except in the rare case where one holds a belief which one acknowledges to be irrational, this would not differ from the totality of propositions which one accepts.

But now, if the total evidence on which we are to make our judgements is to be limited to propositions which we already accept, the fatal objection arises that if one has acquired even a single piece of evidence which is relevant to the hypothesis which one is trying to evaluate, there is no reason why one should bother to look for any more. For instance, if I am betting on the result of a football match between teams A and B, and my total evidence is that team A won on the last occasion that they played each other, I shall be doing all that the theory requires of me if I bet on team A, even though a little further investigation would show me that on current form team B was superior. The obvious way of dealing with this difficulty is to extend the concept of total evidence so that it includes some evidence which is not yet in my possession but is available to me. But the result is that the concept again becomes obscure: for how am I expected to determine what evidence is available to me?

However, I do not want to make too much of this. A methodological principle, considered simply as a guide to conduct, may be expected to be vague: so, let us say that the principle is simply that we should always try to maximise the relevant evidence. In actual practice, this could be taken too far. There are times when it is rational to act on evidence which one knows to be incomplete: a general who refused to launch an attack until he had ascertained the position of every enemy soldier would not be very successful. There may also be moral as well as practical limitations. If I am trying to pick a winner on a race-course, the principle of maximising evidence may be held to fall short of requiring me to attempt to bribe the stable-boys. But, subject to such provisos, the principle is in accord with common sense. To return to my example of the football match, we should all agree that a forecaster who based his opinion on a careful study of current form would be behaving more rationally than one who went by the result of the last meeting between the teams and nothing else.

But how can this policy be upheld in terms of the logical theory which we are discussing? In the ordinary way, if we were asked why it was more rational, in my example, to investigate current form, our answer would be that it would yield the better estimate of the teams' respective chances. But this answer is not available to those who hold that, in this type of instance, 'p is probable' means 'p is probable relatively to q'. An estimate which is based on more, or more varied, or more recent evidence can, on this view, be no better than one which is based on less. If to say that something is likely to happen is just to say that it is confirmed by such and such evidence, then any true statement of this kind is as good as any other. Neither can one escape the argument by talking of the principle of maximising evidence as a methodological principle. For a methodological principle is not selected by caprice: it is supposed to have some utility. Its justification is that by adopting it we shall achieve better results. But what are the better results in this case? The obvious answer is 'better estimates of what is likely to happen'. But this is just the answer which we have seen that the logical theory is debarred from giving.

I first put forward this argument in an essay which was published in 1957,[1] and I have not yet seen any effective rejoinder to it. All that is ever said is that the principle of total evidence is to be accepted as a methodological principle, which, as we have just seen, is not an acceptable answer. Without my knowing it, the same argument had been adumbrated by G. E. Moore, in 1953 or thereabouts, in an entry in his *Commonplace Book*. He concludes, rightly, that there is an absolute sense of words like 'probable', 'highly probable', 'more likely than not', etc., in which when we say that p is probable, we are not saying

[1] 'The Concept of Probability as a Logical Relation', in *Observation and Interpretation*, Proceedings of the Ninth Symposium of the Colston Research Society at Bristol University; reprinted in *The Concept of a Person* (1963).

that p is probable, relatively to q. He says that this sense is 'certainly extremely common and probably far more common than any other',[1] and here again he is right, if he is referring to the use of words like 'probable', as they occur in judgements of credibility. Neither is this simply a question of conformity to ordinary usage. Even if what we meant by saying that p was probable were most commonly that p was probable relatively to q, we should still need some other means of expressing the opinion that one such judgement of credibility was superior to another as an estimate of what was likely to happen.

This is not, of course, to say that judgements of credibility are not based on evidence, or that they are justified otherwise than in terms of the evidence on which they are, or could be, based. The point is only that they are not judgements *about* the evidence on which they are based. But then what sort of judgements are they? What can they be said to be about?

B. HOW JUDGEMENTS OF CREDIBILITY FUNCTION

I am not able to answer these questions directly by offering an analysis of judgements of this kind, but I shall try to elucidate it by saying something about the way in which they function. At one time I was inclined to hold that they were to be explained in terms of their force rather than their content. I took the view, which has also been taken by others, that when I say 'It is probable that p', in the sense which is here in question, I am expressing guarded confidence in the truth of p, encouraging others to feel the same, and implying, though not stating, that I have good reasons for my attitude. But while this performative element is undoubtedly present in statements of this sort, I now think that it is wrong to regard them as being merely or even primarily performative, if only because of the different ways in which they can be mistaken. For not only can I be said to be in error if the evidence on which I am relying is not such as to support a reasonable belief that p, but even when my evidence is up to standard, I can still be said to have been in error if further evidence makes it appear more probable that not-p. Nevertheless, I think that we should be erring in the other direction if we entirely assimilated statements of this kind to statements of fact. If I say, for example, that it is probable that the Republicans

[1] *The Commonplace Book of G. E. Moore*, ed. Casimir Lewy (1962), p. 402.

will win the next presidential election in the United States, my statement is not straightforwardly true or false, in the way it would have
been if I had said, without qualification, that the Republicans were
going to win. And one reason why it is not is that there is no definite
body of evidence on which its truth or falsehood depends. Someone
may produce some statistics which I had not taken into account and
by that means convince me that I was wrong. In the light of this new
evidence I agree with him that it is not probable that they will win.
But this need not settle the matter. It may be that someone else can
show that there are special factors which make these statistics untrust-
worthy, or further evidence may be produced which will again
require me to revise my opinion. Neither is there any theoretical limit
to this process. So long as we are discussing probabilities there is no
point at which it is not open for further evidence to turn the scale.
There may, indeed, come a point at which we take the evidence to be
conclusive, and then, it is interesting to note, we no longer speak in
terms of probability. I do not say that it is probable that the Republicans
won the last election, because it is a matter of common knowledge
that they did. Or rather, I do not say it unless I am making a philosophical point. There are those, like Russell, who do say that every
empirical proposition, other than a purely experiential one, which
monitors a present percept, is at best only probable, just on the ground
that it does not follow from any set of purely experiential propositions.
But then, it is to be remarked, the same considerations apply as in the
case of more ordinary judgements of credibility. If I say, as Russell
would have, that it is no more than very highly probable that there is
a table in front of me, I am admitting that there could be further
evidence which would alter this probability, and I am setting no limit
to its possible extent.

That judgements of credibility are not entirely factual, in the sense
that I have been trying to explain, is brought out by the way in which
we view them retrospectively. To continue with my example, let us
suppose that the Republicans lose the next election. Then someone who
had not agreed that it was probable that they would win, might say to
me: 'You see, you were wrong.' I might then reply: 'Well, I only said
it was *probable* that they would win', and this would be a way of
exculpating myself. I should be reminding him that I had not entirely
committed myself to their winning. I should not, however, be denying
that I had been mistaken. But I could try to deny it. I might say: 'No,
I wasn't wrong at all. I said it was probable that they would win, and it

was probable. I cannot help it if the improbable happened.' Would this answer be acceptable? There is no question but that it would be if judgements of credibility were judgements *about* the evidence on which they were based. Neither is there anything odd about saying of something which did not in fact happen that it was reasonable to believe that it would happen, so long as the evidence that it was not going to happen was not in one's possession and not readily accessible. On the other hand, there is some ground for thinking that the suggestion, which Moore offered in an account of his absolute sense of probability, that 'when you say that something is probable what you are saying is merely something as to what it's reasonable to expect ',[1] though it works well enough for the present tense, breaks down when it comes to the past. When something which we had reason to believe would happen does not happen, we are inclined to say not that it was probable that it would happen, but rather that it *seemed* probable. One reason for this is the strength of our belief in causality. If the event in question did not take place, we assume that there must have been signs that it was not going to, even though they may have been signs which it would not have been reasonable to expect anyone to detect. There is also the fact that, in ordinary usage, probability holds a place for truth, a place which it gives up when truth comes to occupy it. On the other hand, to talk of something as seeming probable is to claim no more than respectability for the judgement that it will occur; and when we know that what was said to be, or to seem, probable did not occur, we are still willing to issue a certificate of respectability to the prediction, but not, any longer, a provisional certificate of truth.

The account which I have been giving of judgements of credibility is, I hope, correct so far as it goes; but it is not the whole story, nor, for our purposes, the most important part of it. For while we are normally more interested in the truth of our judgements than in what I have called their respectability, still, in the cases where we do not know them to be true, their respectability is all we have to go by. But a judgement is respectable, in this sense, just to the extent that one has good reason to accept it. Obviously, then, it is not only judgements of credibility that need to be backed by evidence. This is true of all judgements, with the possible exception of judgements of perception and memory, for which some philosophers would say that the request for evidence was inappropriate. But even in their case there is at least the requirement that they be justified, if not on the basis of other judgements, then in

[1] *The Commonplace Book of G. E. Moore*, p. 403.

terms of the experiences on which they are founded. Thus, there is no particular reason for us to focus our attention on judgements of probability, in any of the senses of the term. For what we are trying to establish is our title to accept any statement of fact, which we have not ascertained to be true.

This brings us back to where we started. For if we have no direct warrant, in observation or memory, for accepting a proposition q, it would seem that if our acceptance of it is not to be entirely arbitrary, it must result from our belief in some other proposition, p, which we do have warrant for accepting. And then the question arises how p and q can be related. This is the question to which our excursion into probability theory has failed to provide a satisfactory answer.

C. GENERALISATIONS OF TENDENCY

Let us begin with the case where p and q are propositions which describe particular states of affairs, and since we have seen that nothing is to be gained by making p entail q, let us assume that they are logically independent. Then if our assumption of the truth of p is to supply us with any reason for accepting q, it must be that we, consciously or unconsciously, accept some empirical generalisation which connects p and q. The connection may be direct, or it may be indirect, by way of a theory from which p and q both follow. In the case where it is direct, the generalisation may take the form of a universal proposition, of which one example would be that whenever an event A, of the type described by p, occurs, an event B, of the type described by q, occurs in such and such a spatio-temporal relation to A, or it may take the weaker form of a statement of tendency. It may state, for example, not that whenever an event of type A occurs, an event of type B occurs in the specified relation to it, but only that this almost always happens, or that it happens more often than not. Many of the generalisations that we rely on, when we reason about human conduct, belong to this class.

Generalisations of this weaker type raise a special problem. If p and q are directly connected by a universal proposition r, p and r will jointly entail q. There will therefore be no question but that if we are justified in accepting p and r, we are justified in accepting q. But if r is only a statement of tendency, the entailment does not hold. From the fact that the event a has occurred and that most often when an event of type *A* occurs an event of type *B* occurs in such and such a relation to *A*, we cannot infer that the relation will hold in this particular instance. Neither does it help us to resort to the proportional syllogism and infer that *b* will probably occur. For, as we have seen, the conclusion of the proportional syllogism, that *b* will probably occur, is no more than a reformulation of the premiss that most often *A*s are accompanied by *B*s. As a statement of frequency, it licenses no conclusion about any individual instance.

In cases of this kind, we have, therefore, to adopt a rule of procedure. The rule is that when two propositions *p* and *q* are linked by a generalisation of tendency and *p* is already accepted, we are to accept *q* with a degree of confidence which is proportionate to the strength of the tendency, unless some special reason can be given why we should not. It is to be noted that I am not here concerned with the establishment of the generalisation of tendency itself. The question of our right to extrapolate from past to future regularities still remains unanswered, and I shall return to it presently. The rule which I am now putting forward bears only on the application of a general statement of tendency, once we are justified, or think that we are justified, in accepting it.

What then is the special reason which, in a case of this kind, should prevent us from accepting *q*, given *p*? Clearly it must consist in our having, or thinking that we have, as good a ground for accepting a stronger generalisation which favours not-*q*. If we accept some universal generalisation and a set of particular statements from the conjunction of which not-*q* follows, this will be decisive. But very often it will be a matter of our having to weigh one generalisation of tendency against another. In that case, the stronger generalisation is not necessarily the one that expresses the more frequent tendency. It is rather the one which takes account of the greater number of factors. The ideal at which we are aiming is to be in a position to apply some universal generalisation from which, in conjunction with our data, either *q* or not-*q* can be derived. If we have to be content with a generalisation of tendency, then indeed, the more frequent the tendency, the nearer the generalisation comes to being universal, the better. But just as in the case of a universal generalisation we have to make sure that the conditions are right, that there are no countervailing factors which will either invalidate the generalisation, in the case where we have overlooked them, or make it inapplicable, in the case where their absence is postulated as a negative condition, so, when it comes to a generalisation of tendency, we have to look for the countervailing factors which would make it even less safe to apply than it is ordinarily. This is the reason why in applying statistical generalisations to the individual instance, we do well to assign the subject to the narrowest class of reference at our disposal. So long as we are concerned only with frequencies, one class of reference must be as good as another, provided that the frequency in either is equally well established. But if our concern is to arrive at least at a probable judgement about what will actually happen, by which, of course, I do *not* mean a judgement about what will probably happen, then it must be our policy to take into account as many relevant factors as we can.

If I am asked to justify this rule, the best answer I can give is that if the guiding principle of the inference from p to q does hold in the majority of cases, then in applying it over its whole range we shall necessarily be right most of the time, and that if we have no reason to treat a particular instance as an exception, we should not do so. This does not mean that I am relying on the proportional syllogism as a justification for accepting the generalisation of tendency. I am concerned here only with the problem of applying such a generalisation, once it has been accepted.

But then what does make it acceptable? Or rather, since their case is more straightforward, and the question not significantly different, let us ask what reason there can be for accepting any universal generalisations. For while there can indeed be no better reason for accepting a proposition, q, than that it is entailed by the conjunction of a true proposition, p, and a true universal proposition, r, it may be objected that we are not in a position to give this as our reason for accepting q, unless we can show that we also have good reason to accept both p and r. But whether we ever can have good reason to accept an empirical generalisation is just what Hume's argument puts in question, and so far we have found no way of meeting it.

D. HEMPEL'S THEORY OF CONFIRMATION

The suggestion which I now wish to examine is that we are given good reason to accept a universal proposition by making observations which confirm it. In discussing this suggestion, we shall need to answer two questions. First, what is it for a generalisation to be confirmed? And secondly, why should the fact that a generalisation has been confirmed be a reason for accepting it? This second question may sound a little strange, but we shall see that it raises a genuine difficulty.

The current answers to the first of these questions take two main forms. One of them, which might be called the orthodox answer, is that we confirm a generalisation by collecting favourable instances of it. The other view, of which Sir Karl Popper is the main protagonist, is that a generalisation is confirmed, or as Popper prefers to say, corroborated, when it resists an attempt to falsify it, the strength of the corroboration being proportionate to the severity of the test. We shall consider later on whether the difference between the two answers is significantly more than a difference of formulation.

One of the best-known expositions of the orthodox view was given by Professor Carl Hempel in his 'Studies in the Logic of Confirmation'.[1] Hempel first introduces the concept of the development of a hypothesis H for a finite class of individuals C, the development being identified with what would be asserted by H if there existed only the objects which are elements of C. With the help of this concept, he then defines what he calls direct confirmation, as follows: 'An observation report B directly confirms a hypothesis H if B entails the development of H for the class of those objects which are mentioned in B.' He then proceeds to define confirmation by saying that 'an observation report B confirms a hypothesis H, if H is entailed by a class of statements each of which is directly confirmed by B'. I shall be concerned here only with the definition of direct confirmation, on which the more general definition depends.

The first point to be made about it is that it applies only to relatively simple hypotheses, of which the factual constituents are either explicitly definable in terms of observational predicates or else conditionally reducible to them.[2] The more abstract hypotheses which figure in many scientific theories would not be covered. It might, however, be possible to extend the scope of the definition by liberalising the notion of development. This could be achieved by identifying the development of a higher-level hypothesis with those of the directly testable hypotheses through which it was verified. The same problem arises, on Professor Popper's theory, in cases where the level of a

---

[1] *Mind*, LIV 213 and 214 (1945).
[2] The allusion is to Carnap's method of defining dispositional terms; cf. his *Testability and Meaning*.

scientific hypothesis is too high for it to be contradicted by any pair of observation statements. One has to decide what is to count as a falsifying instance. For our present purpose, however, these problems can be set aside. If it can be assumed that the confirmation of any hypothesis, of whatever level, depends in the last resort upon observation, we may be content to see how our theories deal with the confirmation of hypotheses which are taken to be directly testable.

According to Hempel, there are four conditions which a definition of confirmation has to satisfy. They are, as he names them, the conditions of Equivalence, Entailment, Consequence and Consistency. The Equivalence condition is that whatever confirms, or disconfirms, one of two equivalent statements also confirms, or disconfirms, the other. The Entailment condition is that any statement which is entailed by an observation statement is confirmed by it, or more generally, that any statement confirms every statement which it entails. The Consequence condition, from which the Equivalence condition actually follows, since equivalent statements are consequences of one another, is that if p confirms q and q entails r, p confirms r. Finally, the Consistency condition is, in Hempel's words that 'every logically consistent observation report is compatible with the class of all the hypotheses which it confirms'.[1]

The entailment and consequence conditions seem to be unexceptionable. Nevertheless, if we do maintain them, we have to reject another principle, which we might otherwise be inclined to adopt. This is the so-called Converse-Consequence condition, according to which, if p entails q, q confirms p. The reason why we have to reject it if we maintain the others, is that they would jointly prove that any statement confirmed any other. For consider any two statements p and q, which need not even be mutually compatible. Then by the converse-consequence condition p would confirm the conjunction of p and q, since it is entailed by it, and by the consequence condition it would confirm whatever the conjunction of p and q entailed, and thereby it would confirm q, whatever statement q might be.

The consequence condition has in any case to be handled carefully. We must not take it as implying that if p entails q, then any addition to the existing evidence which has the effect of more strongly confirming p will also have the effect of more strongly confirming q. For Carnap has shown that there are counter-examples to this, at least if we allow degrees of confirmation to correspond to the proportion of possible

[1] Op. cit., p. 105.

cases which are favourable to the given hypothesis on the evidence in question: in some cases where q is a disjunctive proposition of which p is one of the disjuncts, evidence which increases the probability, in this sense, of p decreases the probability of q, which was previously higher than that of p.[1] The most that can be assumed is that if p entails q, the degree to which a given body of evidence confirms q can never be lower than that to which it confirms p. But even this principle would be violated if we admitted both the consequence and the converse-consequence conditions. For in that case we should have to allow that a body of evidence which was favourable to p, but neutral to q, confirmed the conjunction of p and q; but then, *ex hypothesi*, it would confirm q less strongly than it confirmed the conjunction from which q followed.

It is clear then that we cannot retain both the consequence and the converse-consequence conditions. What is not so clear is that the converse-consequence condition is the one to be sacrificed. The argument in favour of keeping the consequence condition is that it goes with the entailment condition, and that it would be absurd not to regard entailment as a form, and indeed the strongest form, of confirmation. It may be objected that confirmation is not necessarily transitive, but against this one can argue that it must be transitive in the cases where the second arch of the bridge is the relation of entailment. On the other hand, it also seems counter-intuitive to say that we do not confirm the truth of a conjunctive statement by verifying one of its conjuncts, or that we can establish the truth of a disjunction without confirming the truth of any of its members. What Hempel does is maintain the converse-consequence principle in the special case in which a generalisation is held to be confirmed by the observation of favourable instances, while denying that the principle is valid universally. There seems, however, to be no sufficient reason for discriminating in favour of this application of the principle, as against its application to conjunctions or disjunctions. And indeed I shall argue later on that if there is any reason to discriminate, it goes the other way.

Hempel's other two conditions are even more troublesome. This is surprising in the case of the condition of equivalence which one might well, at first sight, be disposed to take for granted. If each of two hypotheses logically entails the other, it would seem reasonable to conclude that they have the same empirical content; and then it would appear very strange that any evidence which confirmed or discon-

---

[1] See Rudolf Carnap, *Logical Foundations of Probability*, part vi, section 69.

firmed either one of them should not equally confirm or disconfirm the other. One might even say that there were not two hypotheses in such a case but two different formulations of the same hypothesis; and then it would appear even more strange that the way in which a hypothesis is formulated should determine whether or not a given piece of evidence confirms it.

E. THE INFAMOUS RAVENS[1]

Nevertheless, as Hempel has shown, the maintenance of the equivalence condition does have a consequence which it is very difficult to accept. It leads to his notorious paradox of the black ravens. If, following Russell, we construe the universal proposition 'All ravens are black' as stating that for all x, if x is a raven, x is black, then, on Hempel's theory, it will be confirmed by anything which is both a raven and black and refuted by anything which is a raven and not black. Let us call this proposition p_1. But p_1 is equivalent by contra-position to the proposition p_2, 'For all x, if x is not black x is not a raven', which is confirmed by any non-black thing which is also not a raven. Moreover, p_1 is also equivalent to p_3, 'For all x, if either x is a raven or x is not a raven, then either x is not a raven or x is black'. Since the antecedent of p_3, being a tautology, is satisfied by everything, it follows that p_3 is confirmed by anything which satisfies its consequent, that is to say, by anything which is either not a raven or black. The result, if we maintain the equivalence condition, is that the hypothesis that all ravens are black is confirmed by the existence of any object which is either a raven and black, or not a raven and not black, or not a raven and black, and is refuted by the existence of any object which is a raven and not black.

If we generalise this result, we arrive at the conclusion, which does seem paradoxical, that in the case of a hypothesis of the form 'for all x, if fx then gx', where the presence or absence of f and g is detectable by observation, there is no such thing as neutral evidence. Every observable fact is relevant. A case of f and not-g refutes the hypothesis, and cases of f and g, not-f and not-g, and not-f and g confirm it. But these are all the cases that there can be. This is not, indeed, a paradox in the sense

[1] The phrase is due to Nelson Goodman. See his *Fact, Fiction and Forecast*, 2nd ed. (1965) p. 70.

that it yields a contradiction, but it does run counter to what we naturally think. We might not be so strict as to maintain that only the discovery of black ravens can confirm the hypothesis that all ravens are black – for instance, the pigmentation of birds of other species might be held to be relevant – but certainly we should not ordinarily take this hypothesis to be confirmed by the fact that I have a white handkerchief in my pocket, or a black fountain-pen. Such facts as these, we should say, are obviously irrelevant.

At this point, there are four courses that are open to us. We can reject the equivalence condition. We can try to show that this condition need not operate in such a way as to produce the paradoxical conclusion. We can try to show that the conclusion is not really paradoxical, in which case we have to explain why it seems to be so. And finally we can deny that a generalisation is confirmed by every instance which satisfies its antecedent and its consequent.

Of these courses Hempel himself favours the third. He does not consider the first course, since he takes the principle of equivalence to be unassailable, nor the fourth, which would imply the rejection of his theory. Concerning the second course, he remarks that it would be possible to interpret the propositions which he uses to illustrate his paradox in such a way that they were not equivalent. This would be achieved by construing universal propositions as implying that their antecedents were satisfied. So the proposition that all ravens are black would carry the existential rider that there are ravens, whereas the proposition that all non-black things are non-ravens would carry the different existential rider that there are non-black things. Consequently, the two propositions would not be equivalent.

Hempel's objections to this proposal are first that it breaks with scientific custom, secondly, that we do not want to be put in the position of having to reject hypotheses about ideal entities, such as bodies on which no forces are acting, just because their antecedents are not satisfied, and thirdly, that in many cases it will not be clear what the existential conjunction should be taken to be: an example which Hempel gives is that of the hypothesis that persons who respond with a positive skin reaction to injections of a certain test substance have diphtheria. Is this to be taken as implying the existence of persons who are injected and have the reaction, or merely the existence of persons who are injected, or merely the existence of persons? And is it plausible to say that in making one or other of these arbitrary choices we are giving the hypothesis a different content?

A point which Hempel does not notice is that the proposal to which he is objecting is anyhow ineffective as a means of avoiding his paradox. What it does is turn universal propositions into conjunctions, which would be confirmed by the confirmation of each of their conjuncts. But then, once we have confirmed the existential conjunct, and turn to the remainder of the hypothesis, we are in the same position as before.

A variant of the second course, which Hempel also discusses, is the suggestion that universal propositions be construed not as containing any existential conjunct but as being restricted in their field of application. Thus the proposition that all ravens are black would be understood as applying only to ravens, and would therefore not be confirmed by the existence of anything other than a raven. This would not mean giving up the condition of equivalence, since the propositions which are equivalent to the proposition that all ravens are black would be regarded as being subject to the same restriction. On the other hand, if one were confronted with the proposition that all non-black things are non-ravens, one might understand it as applying only to non-black things, and in that case one would have to take the proposition that all ravens are black, to which it is equivalent, as being confirmed not by the existence of black ravens, but only by the existence of objects which are neither black nor ravens. Professor Israel Scheffler, in the illuminating discussion of the paradoxes of confirmation which he has included in his book *The Anatomy of Inquiry*, argues that to adopt this proposal would be to take, not the second, but the fourth of the courses which I listed, that of giving up the principle that a generalisation is confirmed by every instance which satisfies its antecedent and its consequent, but here I think that he is mistaken. It is true that if the proposition that all ravens are black and its equivalents are understood as applying only to ravens, the existence of something which is neither black nor a raven cannot be taken as confirming the proposition that all non-black things are non-ravens, but the reason for this is that when the generalisation that all non-black things are non-ravens is interpreted in this way, the instances which appear to satisfy its consequent do not really do so, since, not being instances of ravens, they fall outside its range. The same would apply to instances which ostensibly satisfied the antecedent and consequent of the generalisation that all ravens are black, if this proposition and its equivalents were understood to refer only to non-black things. It is because this proposal, while not denying the equivalence of propositions like 'All ravens are black' to their contra-positives, does differentiate between a hypothesis, relating only to ravens, which would be most naturally expressed in this form, and a hypothesis, relating only to non-black things, which would be most naturally expressed in the form of the contra-positive, that I take it to be a variant of my second course.

Though it escapes some of the objections on which its predecessor foundered, this proposal also has the defects that it interprets universal propositions in a way that is not in accord with scientific practice and that in many cases one will be uncertain what the range of a generalisation should be understood to be. But Hempel's main objection to it is that it is undesirably restrictive. He takes as an example the hypothesis that sodium salt burns yellow. One use to which this hypothesis can be put, if one accepts it, is to determine whether some unidentified substance is sodium salt: if it fails to burn yellow, one can infer that it is not. But this use of the hypothesis, as a negative test, would not be legitimate if it were applicable only to substances that were sodium salt. Admittedly, the hypothesis could be construed as applying only to things that did not burn yellow, but then its use would be illegitimate if the result of the test were positive. In either case, there is the difficulty that the result of the test has to be known before it can be determined whether the hypothesis is equipped to make it. I agree, therefore, with Hempel that this proposal is not 'attractive. If we reject it and keep the equivalence condition, we must either tolerate Hempel's paradox, or else find some argument to show that for a generalisation to be confirmed on a given occasion it is not always sufficient that it be satisfied.

Hempel himself chooses to tolerate the paradox. He maintains that the hypothesis that all ravens are black really is confirmed by the existence of anything whatsoever, other than a non-black raven, and tries to explain why we mistakenly regard this view as paradoxical.

He gives two reasons. The first is that we misconceive the scope of universal propositions. Because it is our interest in their grammatical subjects that leads us to formulate such hypotheses as that sodium salt burns yellow or that all ravens are black, we take their grammatical subjects to be all that they are about. But this is a mistake. From a logical point of view, the scope of these propositions is unlimited: they are about everything that there is. For what they do is deny that certain properties are conjoined in nature. In the examples chosen, what is denied is that anything at all has the properties of being sodium salt and not being yellow, or being a raven and not being black. So every object either conforms to such a hypothesis or violates it.

The second reason which Hempel gives for our reluctance to admit that the existence of things which are not *A* can confirm the generalisation that all *A*s are *B* is that we are influenced by our previous knowledge. If someone attempted to prove to us that sodium salt burned yellow by holding a piece of ice to a flame and showing that it did not burn yellow, we should find his procedure strange. What is that to the purpose? we should say. But the reason why his experiment would seem to us irrelevant is that we already know that ice does not contain sodium salt, so that we are sure in advance that the experiment will not yield a counter-example to his hypothesis. On the other hand, if we did not know that what was being held to the flame was a piece of ice, or if we did not know that ice did not contain sodium salt, the experiment would not be irrelevant. To discover first that the unknown substance did not burn yellow and then that it was not sodium salt might not be the most natural way of setting about confirming the hypothesis that sodium salt burns yellow; but it would be a legitimate procedure.

I think that there is something in these arguments. Whether or not universal propositions commonly are construed as applying to everything that there is, the recommendation that they should be seems to be acceptable; and it may also be conceded that in cases where we have cause to think that we may have come across a counter-example to the generalisation that all *A*s are *B*, the discovery that this thing which is not *B* is also not an *A* may reasonably sustain our confidence in the generalisation. Nevertheless Hempel's second argument misses the main point. What seems paradoxical is that the generalisation that all *A*s are *B* should be confirmed by the mere existence of something which is not an *A* no less than by the existence of something which is both an *A* and a *B*; and this seems equally paradoxical whether or not the thing which is not an *A* is already known to us not to be one.

I think that what our intuition requires is that the existence of such things as a white handkerchief or a black fountain-pen should be regarded as having no relevance at all to the proposition that all ravens are black, but it would be somewhat appeased if evidence of this kind was at least not put on a level with that of the existence of black ravens. A suggestion which would have this effect has been made by Janina Hosiasson, in an article 'On Confirmation' which appeared in the *Journal of Symbolic Logic* in 1940 and by David Pears in an article on 'Hypotheticals' which appeared in *Analysis* in 1950. The point from which they start is that one may reasonably believe the class of ravens to be very small relatively to the class of non-black things and even smaller relatively to the class of things which are not ravens. Now the hypothesis that all ravens are black can be fully established either by examining everything that there is and failing to find any non-black ravens, or by examining all the non-black things and finding no ravens among them, or by examining all the ravens and finding them all black. But, the relative size of these classes being what it is, the third procedure is much the most economical. And not only that, but it is also more productive. The same number of observations will represent a higher proportion of our total task. If the world were such that it contained a great many ravens and very few things that were not black, the sensible and more productive course would be to pick out the non-black things and see if any of them were ravens.

The same argument can be put in another way. If the hypothesis that all ravens are black is to be refuted, the counter-examples must occur among the things discovered to be ravens one or more of which turns out not to be black, or among the things discovered not to be black, one or more of which turns out to be a raven. But since the class of ravens, which we are assuming to be finite, is smaller than that of non-black things, the discovery of a black raven carries more weight in the scale of confirmation than the discovery of a non-black non-raven. It carries more weight because it diminishes in a greater proportion the number of possible counter-examples: there remain fewer ravens with a chance of not being black than there remain non-black things with a chance of being ravens.

If we generalise this argument, we find that it rests upon the following principle. Let H be a hypothesis of the form 'For all x, if fx then gx'; let U be the spatio-temporal region over which it is understood to range; let E_1 and E_2 be non-identical states of affairs within U, each of which is compatible with H; and let us say that a state of affairs E threatens a hypothesis H if E exhibits at least one of the properties whose conjunction is denied by H. Then for E_1 to confirm H more strongly than E_2 does, it is sufficient that either E_1 threatens H and E_2 does not, or that they both threaten H, but the number of states of affairs within U which threaten H in the way E_1 does is smaller than the number of those which threaten H in the way E_2 does.

But now, having elicited this principle, I must say that I cannot see any logical justification for it. If all that is at stake in our example is the non-existence of anything which combines the property of being a raven and not being black, then I cannot see why a white handkerchief

THE PROBLEM OF CONFIRMATION 73

should not be as good a confirming instance as a black raven. Indeed, it might well be argued that it was a better instance: unlike the black raven, it does not even encroach on the forbidden territory. That the higher marks should go to the more dangerous instance would seem to be contrary to merit, like the fatted calf going to the prodigal son.

What makes the principle seem plausible is that in searching for a counter-example to a hypothesis of the kind in question, one thinks of all the objects which fail to satisfy its antecedent as already having been eliminated, and therefore of an object which does satisfy both the antecedent and the consequent as increasing, beyond what is already a high figure, the probability that no counter-example exists. But from this point of view no instance counts for more than any other in building up the probability. One would have to seek a psychological explanation for the fact that it is only at a certain stage in this process that the confirming instances are taken to be significant.

In any case, even if we accept this principle, we are not freed from Hempel's paradox. For, as both Hempel and Scheffler have pointed out, we still encounter it in cases where we have no reason to believe that fewer things satisfy the antecedent of a hypothesis than fail to satisfy its consequent. One of Scheffler's examples is the hypothesis that all invertebrates lack kidneys.[1] There are not known to be fewer invertebrates than there are animals that have kidneys; yet it would still seem paradoxical to find confirmation of this hypothesis in the fact that one's incontinent dog is vertebrate.

A further objection which Scheffler raises[2] is that even in the more favourable example of the ravens, the suggestion which we are now examining goes against our intuition. For since it maintains the condition of equivalence, it requires that an instance of a black raven should count for more than an instance of a non-black non-raven not only as confirming the hypothesis that all ravens are black but equally as confirming the hypothesis that all non-black things are non-ravens. Yet we intuitively feel in the case of both hypotheses that the best way to confirm them is to find instances which satisfy their antecedents and their consequents. Of course our intuition may be at fault; but this has not yet been shown; neither have we yet been able to explain why it operates as it does.

[1] Cf. I. Scheffler, *The Anatomy of Inquiry*, p. 284.
[2] Ibid., pp. 284–6.

F. POPPER AND THE RAVENS

The failure of this attempt to neutralise Hempel's paradox has a bearing
also on Professor Popper's theory of confirmation. It might appear at
first sight that there was no problem here for Popper, since he explicitly
rejects what he calls inductivism: if a theory of confirmation is meant
to show how we can be justified in accepting hypotheses on the basis
of evidence which does not entail them, then Popper would say that
he does not have such a theory and that those who do are pursuing a
will-o'-the-wisp. Nevertheless, Popper does think that hypotheses
require to be tested, if possible under the conditions in which we think
that there is the greatest chance of their being falsified; and when they
survive our attempts to falsify them, they are said to be corroborated.
But now it is very hard to see how anything different can be meant by
saying that a hypothesis has been corroborated, in this fashion, from
what the inductivists mean when they say that a hypothesis has been
confirmed. For what would be the point of testing a hypothesis, if it
gained no credit from passing the test? Why even should we abandon
a hypothesis which had failed a test unless we thought that this proved
it to be unreliable? Admittedly, a generalisation to which an exception
has been found is shown not to be universally true, but we could regard
its failure as having no bearing on its future career; we could even
believe that a hypothesis which had been falsified was the more likely
to hold good in future cases. As I once put it,[1] falsification might be
regarded as a sort of infantile disease which even the healthiest hypotheses could be depended on to catch: when they had once had it,
there would be a smaller chance of their catching it again. If we do not
proceed in this way, it is because we think that a hypothesis which has
failed once is likely to fail again; and conversely, the more tests a
hypothesis passes, or at any rate the more tests of what we take to be
the proper kind, the more confidence do we think that we can have in
its continuing to be successful. Popper himself appears to concede this
when he says, in the course of criticising Carnap, that 'the better a
statement can be tested, the better it can be confirmed, i.e. attested by
its tests'.[2]

But if what attests a hypothesis is only that our observations fail to
refute it, then again it is not clear why some of its successes should

---

[1] *The Problem of Knowledge*, p. 79.
[2] K. Popper, *Conjectures and Refutations*, p. 267.

count for more than others. We return to the problem why a state of affairs which threatens a hypothesis by satisfying its antecedent should carry more weight than one that threatens it by failing to satisfy its consequent, and why either should carry more weight than one that does not threaten it at all. An answer which Professor Watkins has given is that a hypothesis is confirmed only by observations which are designed to test it,[1] and commonly we require to know that a state of affairs satisfies the antecedent of a hypothesis before we are willing to treat it as a test. But what of the case in which we unexpectedly come upon a state of affairs which falsifies a hypothesis which we are not engaged in testing? Presumably Watkins would not wish to say that the hypothesis remained unviolated, just because its violation had not been sought; and if he did he would clearly be wrong. But then if an observation is allowed to count against a hypothesis, irrespectively of our intention, it is not clear why in the event of its being favourable to the hypothesis it should be debarred from counting for it. Moreover, even if this arbitrary rule were adopted, it would not meet the diffi-culty. For suppose that someone did set out to test the hypothesis that all ravens are black simply by accumulating instances of anything other than non-black ravens. On what grounds could this procedure be held to be irrational?

I think that a follower of Popper would have to say that what made this man's procedure irrational was that the instances which he collected confirmed the hypothesis only very weakly. For stronger confirmation, he would have to rely on instances which faced the hypothesis with a severer test, whether or not they were designed to do so. But how are we to measure the severity of a test? The answer would appear to be that, of two instances which conform to a given hypo-thesis, one constitutes a severer test of it than the other if there is antecedently a greater probability of its being a counter-example. But how is this to be decided? Why, for instance, in the case of a black raven and a white handkerchief, is it antecedently more probable that the raven should turn out not to be black than that the non-black thing should turn out to be a raven? Is it because of the relative difference in number, so that non-black ravens, if there were any, would constitute a higher proportion of ravens than they would of non-black things? But if this is the answer, we are back in the position which we reached in our discussion of Hempel's theory and the same objections arise.

Not only that, but the emphasis on falsification, which distinguishes

Popper's theory from Hempel's, creates a special difficulty. If hypotheses were accredited only through the failure of our attempts to
falsify them, then in testing a hypothesis of the form 'for all *x*, if *fx*,
gx', we ought to lose interest in the result as soon as we discover the
presence of *g*. For since the hypothesis can be falsified only by the
concomitance of *f* and not-*g*, nothing that has *g* can be a counter-
example to it. But this would not be scientific procedure, as we can see
by returning to Hempel's example of the sodium salt. Suppose that we
are experimenting with a substance of which we do not yet know
whether it contains sodium salt or not, and that we find that it burns
yellow. If all that mattered was that the experiment should not falsify
the hypothesis that sodium salt burns yellow, there would be no point
in proceeding with it any further. The mere fact that the substance did
burn yellow, whatever its composition, would ensure that the hypothesis was not meeting with a counter-example. But in practice we
should proceed with the experiment. We should find out whether the
substance did contain sodium salt, and if it did, we should regard the
hypothesis as having been confirmed, in a way that it would not have
been confirmed if its antecedent had not been satisfied.

G. GOODMAN'S REJECTION OF THE EQUIVALENCE CONDITION

Though I do not remember seeing it advanced as a special objection to
Popper's theory, there is nothing new in what I have just said. It again
brings up the point that not every state of affairs which conforms to a
given hypothesis, in the sense that it fails to violate it, is thought to
confirm the hypothesis in an equal degree. But then we return to
Hempel's paradox, which we have not yet found a way of showing
either to be avoidable or to be innocuous.

At this point it may be worth taking another look at the equivalence
condition from which the paradox is derived. Can it be rejected? The
case for maintaining it is, indeed, very strong. One of the criteria for
deciding that two sentences are synonymous is that the propositions
which they express are confirmed to the same degree by the same
evidence, and it is at least arguable that when two propositions are
logically equivalent the sentences which express them may be held to
be synonymous. Even if this is not admitted, it can hardly be denied
that logically equivalent propositions come out true or false in the

same empirical situations, and in that case it is not easy to see how evidence which is relevant to one of them could fail to be relevant to the other. Nevertheless, if we were in a position where we had no other resource than to reject the equivalence condition, there is a line that we could follow. Taking up a suggestion which was first made by Professor Nelson Goodman in his *Fact, Fiction and Forecast*, we could maintain that a necessary condition for anything to confirm a proposition of the form 'All *A*s are *B*s' was that it did not also satisfy the proposition that no *A*s are *B*. This would allow an instance of a black raven to confirm the generalisation that all ravens are black, but not that of a white handkerchief or a black fountain-pen, since, by not being black ravens, they also satisfy the generalisation that no ravens are black. The white handkerchief would still confirm the generalisation that all non-black things are non-ravens, but the black raven and the black fountain-pen would not, since they would also satisfy the generalisation that no non-black things are non-ravens. All three would confirm the third of our equivalent generalisations – that everything is either black or not a raven, since they all violate the hypothesis that nothing is either black or not a raven, but that is not paradoxical. What is suggested here is that we take advantage of the logical fact that whereas the propositions 'All ravens are black', 'All non-black things are non-ravens' and 'Everything is either black or not a raven' are logically equivalent, their respective contraries, in the Aristotelian square of opposition, 'No ravens are black', 'No non-black things are non-ravens' and 'Nothing is either black or not a raven' are very far from being so.

The rule which we are here being encouraged to adopt is not, as one might at first suppose, that no evidence confirms a hypothesis if it also satisfies its contraries.[1] For in the first place, the propositions that all ravens are black and that no ravens are black are not strictly contrary, since they would both be true if there were no ravens; and secondly, we shall see later on that Goodman himself has an argument to show that given any body of evidence which confirms a hypothesis of the form 'All *A* is *B*', we are always able to formulate a contrary hypothesis which is equally confirmed by it. The rule is rather that no evidence is to be taken as confirming a hypothesis of this form, if it also satisfies the hypothesis that no *A*s exist.

The effect of this principle, evidently, would be to confine the

---

[1] Cf. Sidney Morgenbesser 'Goodman on the Ravens', *Journal of Philosophy*, LIX 18 (1962).

favourable instances of a universal hypothesis to those that satisfied its antecedent. It could, therefore, be charged with being too restrictive. Like the proposal to make the satisfaction of their antecedents a limiting condition on the range of such hypotheses, it would rule out negative tests. We should not confirm the hypothesis that sodium salt burns yellow by finding that some substance which did not burn yellow was not sodium salt. This is, indeed, an undesirable but not, perhaps, a fatal restriction, and it could be a price worth paying for the elimination of Hempel's paradox. What is more doubtful is whether we should go so far as to sacrifice the equivalence condition, in what would appear to be an arbitrary fashion. I shall try to show in a moment that we can reach a position very similar to Goodman's at a smaller cost.

H. QUINE'S ATTEMPT TO SOLVE HEMPEL'S PARADOX

Of the four courses which I listed, the only one that we have not so far discussed is that of denying that a generalisation is confirmed by every state of affairs which satisfies its antecedent and its consequent. This is the line taken by Professor Quine, in an essay on 'Natural Kinds', which is reprinted in his book on *Ontological Relativity*. He there suggests that while an instance of a black raven does count in favour of the generalisation that all ravens are black, and thereby indirectly in favour of the generalisation that all non-black things are non-ravens, an instance of a green leaf does not count in favour of the generalisation that all ravens are black, because it does not count in favour of the generalisation that all non-black things are non-ravens either. As Quine puts it, 'being a non-black thing' is not a projectible predicate, with the result that its instances do not confirm any generalisation. Quine does not say how we are to decide when a predicate is projectible, but he thinks that it may be true that when a predicate can be projected, its complement can not be; and this would be enough to frustrate Hempel's paradox.

Unfortunately, the principle which Quine needs for this result does not appear to be acceptable. Consider an example which Professor Max Black has given in his 'Notes on the Paradoxes of Confirmation', which appeared in a collection called *Aspects of Inductive Logic*.[1] The example is that of the generalisation 'All vertebrate animals are warm-

---

[1] *Aspects of Inductive Logic*, ed. J. Hintikke and P. Suppes (1966).

blooded animals' and its contrapositive 'All cold-blooded animals are invertebrate animals'. Surely the predicates 'warm-blooded animals' and 'cold-blooded animals' are equally projectible. Quine has, indeed, sought to provide against this kind of example by saying that 'we must not be misled by limited or relative complementation'.[1] But then all we have to do to restore Hempel's paradox is to substitute 'All ravens are black animals' for 'All ravens are black'. This does, indeed, do away with the white handkerchief and the black fountain-pen, as confirming the blackness of ravens, but it leaves us with white mice and black cats which from this point of view are no less undesirable.

In case it is objected that the predicate of being a non-black animal is no more projectible than that of being non-black, let us take another example, which also has the advantage, I believe, of setting the whole problem in a clearer light. Consider the proposition that all the seats in a particular railway compartment are reserved. Let the compartment in question be compartment *A*, and let the proof that a seat is reserved be that it has a ticket on it saying so. Then in what way does the observation that one of the seats in compartment *A* is ticketed confirm the proposition that they all are? The answer is not that when we find that one of the seats has been reserved, we are thereby given any reason to expect that the others will be. If what is meant by saying that a predicate is projectible is that the fact of its being satisfied by an object increases the probability, in my third sense, of its being satisfied by other objects of the same kind, then neither the predicate of being reserved nor that of being unreserved is, in general, projectible. This is not to say, however, that the generalisation that all the seats in compartment *A* are reserved is not confirmed by observing that one of them is. It is confirmed for the simple reason that in order that the generalisation should be true, it is necessary and sufficient that each of a set of singular propositions be true, and in observing that one of the seats is reserved, we establish the truth of one of the members of this set.

But now suppose that we find someone examining the other compartments of the train, or even just wandering about the station, and that when we ask him what he is doing he tells us that he too is trying to confirm the proposition that all the seats in compartment *A* are ticketed. He is going about it indirectly by confirming the equivalent proposition that anything which is not ticketed is not a seat in compartment *A*. Clearly his behaviour strikes us as ridiculous, but is this simply because he must be wasting his time, since it is obvious that

[1] *Ontological Relativity and Other Essays*, p. 116.

any counter-example to the proposition which he says that he is testing will if it exists at all, be found inside compartment *A*, or is there more to it than that?

I think that there is something more to it. It is not just that our man is going about his business in the wrong way; he is going about the wrong business. He is confirming the wrong generalisation. But how can it be the wrong generalisation if it is equivalent to the one that he wishes to confirm? The answer is that, for his purpose, they are not equivalent. The proposition that all the seats in compartment *A* are reserved is a summative generalisation. It is not, indeed, logically equivalent to the conjunctive proposition that seat number 1 is reserved and seat number 2 is reserved and so forth, since it does not enumerate the instances which it collects, but it operates as a conjunction, in that it summarises a finite number of items of information, which, even if they are not individually listed, are pin-pointed as falling within a designated area. On the other hand, the proposition that everything which is not ticketed is not a seat in compartment *A* is not a summative but an open generalisation; and if we turn it into a summative generalisation by restricting its range, say, to all the seats in the train of which compartment *A* is part, the propositions which it summarises are not identical with, though they include, the propositions which are summarised by the generalisation that all the seats in compartment *A* are ticketed.

I. AN ATTEMPT TO BETTER QUINE'S SOLUTION

Now I suggest that the belief that we confirm a generalisation of the form 'For all *x*, if *fx* then *gx*', by finding things which are cases of *f* and *g*, but not by finding things which are not cases of *f*, is due to our tacitly counting such generalisations as conjunctions. The process of confirmation then consists in working through the conjuncts. When one of them is verified, the credit of the generalisation is strengthened, not in the sense that the remainder of the race which it has to run becomes any easier, but just in the sense that there is one fewer obstacle at which it can come to grief. This does not apply to an open generalisation since, its range being infinite, there can be no question of our working through its instances. In this way we can find some justification for Carnap's saying that the degree of confirmation of every universal proposition, in an infinite universe, always remains at zero. This is, indeed, objectionable if one equates degrees of confirmation with probability, as Carnap apparently does, and allows oneself to draw the inference that it is as good as certain that to every universal proposition of this kind there is at least one counter-example. But all that we need take it to imply is the necessary proposition that an infinite set is not diminished by the subtraction of any finite number from it; and from the fact that a universal proposition remains unconfirmed in this sense, we can of course infer nothing at all about the likelihood of there being any counter-example to it.

But surely we do not want to say that an open generalisation can not be confirmed to any degree at all? The proposition that all ravens are black is perhaps not a very good example of an open generalisation, but if it is so construed, as it can be, then surely it must acquire some credit from the fact that a number of ravens are discovered to be black and none not to be. The answer to this, I think, is that it does acquire credit but in a different way. What is confirmed by the discovery of *n* black ravens, and none not-black, is the truth of any consistent finite set of propositions which includes the sequence of propositions stating that these particular ravens are black. The total conjunction is confirmed just in the sense that part of it has been successfully worked through. But in confirming this part of the conjunction one is also confirming the security of the inference from being a raven to being black in this set of instances. The open generalisation 'For all *x*, if *x* is a raven *x* is black' then acquires credit from the fact that these inferences exemplify a pattern to which it gives a general licence. This is not to deny that open generalisations have a truth-value. The point is just that they can be employed as rules of inference, and that it is only in this capacity that they are open to confirmation.

But why should not the discovery of a number of white handkerchiefs serve equally well as confirming the inference from not being black to not being a raven? Here I think that Quine has the right answer. The existence of the white handkerchiefs would indeed confirm a summative generalisation which predicated of a list of non-black objects, in which they were included, that they were not ravens, but it does not confirm any rule of inference which proceeds from the property of not being black, because we do not operate with any such rule. This is what Quine meant by saying that the property of not being black was not projectible. It is true that the validity of an inference from f to g carries with it the validity of the inference from not-g to not-*f*, but we have to consider in what direction the inference originally proceeds. In the case under consideration, the inference either from being a raven to being black or from not being black to not being a raven is sustained by a rule concerning the properties of ravens, but not by a rule concerning the properties of non-black things.

If this is right, the solution to Hempel's paradox is that if the propositions that all ravens are black and that all non-black things are non-ravens are treated as summative generalisations they are not equivalent, and that if they are treated as open generalisations, they sustain only a single rule of inference which proceeds in the direction from being a raven to being black and not the other way around. Why this should be so is a complicated question. All that I can say here is that it depends upon our beliefs about the way the world is organised.

J. GOODMAN'S PARADOX

But now we come upon a further difficulty, which is that the process of confirming a generalisation by running successfully through its instances does not in itself justify any inductive argument. It has yet to be shown that the fact that out of conjunction of *n* propositions, *n*-1 have been found to be true, gives us any reason to suppose that the remaining ones will also be true. And indeed, in my example of the reserved seats, there would be no good reason to suppose this, unless we had some further information such as that compartments are usually reserved *en bloc*.

This difficulty comes out more clearly when one looks at the fourth of Hempel's conditions, the consistency condition, which, as we have seen,[1] he expresses in the form: Every logically consistent observation-report is logically compatible with the class of hypotheses which it confirms. This formulation is not very clear, but if it is meant to imply that one and the same observation-report can not be favourable to incompatible hypotheses, it is far too rigorous. So much so, indeed, that it may well be universally violated, in that in every case in which a set of instances exemplifies a hypothesis *H*, it is possible to find another hypothesis *H'*, incompatible with *H*, which they equally exemplify. In the case of quantitative hypotheses, this is illustrated by the fact that any finite series of points will lie on an indefinite number of different

---

[1] See above, p. 65

lines. In the case of qualitative hypotheses, it is illustrated by the possibility of using predicates of the sort that Nelson Goodman introduced in his *Fact, Fiction and Forecast*. The stock example is that of the predicate 'grue' which is meant to apply to anything which is either green and examined before some arbitrarily chosen time *T*, or blue if examined after *T*. Then any evidence, acquired before *T*, which exemplifies the hypothesis that all emeralds are green also exemplifies the incompatible hypothesis that all emeralds are grue.

There has been a good deal of argument about Goodman's predicates, most of it centring on the question whether they are positional. It is easy to show that they can be introduced in a way that does not involve any explicit reference to a particular point in time; and it can also be argued, as Goodman himself has argued, that whether or not a predicate is positional depends on what predicates are taken as primitive. If we also introduce the predicate 'bleen' which applies to anything which is blue if examined before *T* or green if examined after *T*, and take 'grue' and 'bleen' as primitive, then 'blue' and 'green' become positional. There is, however, still a case for saying that 'grue' and 'bleen' are not phenomenal predicates in the way that 'blue' and 'green' are, though I am not sure how much turns on this. I shall not here enter into this dispute, principally because I think that it misses the main point.

The main point, as I see it, is that what Goodman is here doing is to cast a hypothesis which we should naturally express by saying that some *A* is *B* and some *A* is not-*B* into a universal form. To bring in an explicit or covert reference to the time at which the instance is examined is one way of doing this, but not the only one. For example, we could introduce the predicate 'greemin' which applies to any instances of a species of object all the instances of which are green, with one exception which is blue. Then the observations of green emeralds which confirm the generalisation 'All emeralds are green' also confirm the incompatible generalisation 'All emeralds are greemin'. Admittedly, 'greemin' is a highly disreputable predicate. It is parasitical on others, in as much as, in order to apply it, we have first to determine which objects form a species; and even then we cannot tell whether it applies to any individual until we have found out whether it applies to all the other instances of the species to which the individual is assigned. Nevertheless, it does the work required of it, unless we take steps to exclude all predicates of this sort by special legislation. For all I know, all emeralds *are* greemin.

The moral which Goodman draws from this is that before we try to operate with any criteria of confirmation, like Hempel's, we have to decide what kinds of predicates we are going to admit into the generalisation. In other words, we have to decide what predicates are projectible. And here it would seem that the choice must be, to some extent, arbitrary. Goodman himself opts for those predicates which are, as he puts it, entrenched, or, in other words, those to which we have become accustomed. It may be remarked how closely this corresponds to Hume's idea of inductive reasoning as the exercise of a natural habit.

But it is not just a matter of our choice of predicates. Even if we are operating only with the most respectable, the most firmly entrenched predicates, a given body of evidence is still going to be consistent with mutually incompatible hypotheses. If there are n emeralds in all, of which just m have been examined and found without exception to be green, the evidence is indeed consistent with the hypothesis that all of them are green, but it is equally consistent with the hypothesis that the property of being green is distributed among them in any other ratio from $\frac{m}{n}$ to $\frac{n-1}{n}$. Hempel makes it easy for himself to disregard these other hypotheses by so framing his account of confirmation that it applies only to universal generalisations, and, as we have just seen, it is in order to get such hypotheses into the form of universal generalisations that Goodman has to bring in his artificial predicates. But no reason is given for this restriction. Unless we make a special ruling for the purpose, there is no evident ground for holding that the observation of our m green emeralds confirms the hypothesis that all n emeralds are green any more strongly than it confirms any of the other hypotheses with which it is consistent.

K. HAS POPPER AN ANSWER TO GOODMAN?

Do we avoid this difficulty by taking Popper's approach? It might be argued that we did on the ground that Popper is not similarly committed to thinking of the evidence as building up support for any one hypothesis at the expense of its rivals, so long as they are all still in the field. For Popper, as we have seen, there is no question of our having to justify our belief in any hypothesis which has not been falsified. There may be considerations, like that of simplicity, which should guide us in our choice of hypotheses, but they do not make the hypotheses which they favour any more likely to be true than any other hypotheses which are consonant with the existing evidence. On the contrary, since we are encouraged to prefer the most powerful hypotheses, they are, in one sense, less likely to be true, in as much as the more powerful of two hypotheses is the one which is exposed to the greater range of possible counter-instances. So we are supposed to select an hypothesis, do all that we can to test it, and adhere to it so long as it has not been falsified. But is there any reason why we should do this? Let us return to our example of the man who does not have confidence in the application of a hypothesis to further instances until it has been immunised by encountering a certain number of exceptions. What could a follower of Popper say to such a man, in order to show him the error of his ways? Nothing, except that he was not abiding by the rules of the scientific game. But this only proves him unorthodox 'by apostolic blows and knocks'. It does not prove him wrong.

But then what would prove him wrong? That he gets punished by the facts? Having bargained for there being just so many exceptions to a generalisation, he finds that there are more. The superiority of our procedure is shown by the fact that we come to grief less often. A greater proportion of our predictions are successful. But even if this is so, we are only able to triumph over an opponent in retrospect. It has not been shown that our procedure will continue to be more successful than his. If he takes the line that the more he has been punished by the facts, the safer he will be in the future, whereas our punishment is still in store, we have nothing to say to him. It is indeed true, as I have shown elsewhere,[1] that there are limits to the possibility of pursuing what one might call a counter-inductive policy: if there are to be people who can adopt such a policy and things to which it can be applied, a certain number of properties must be constantly combined. But these limits still leave room for a considerable range of different procedures.

L. CRITERIA OF EVIDENCE

What we should like would be a proof that our actual methods of projection are at least the most rational. But I think that it has become clear by now that any attempt at such a proof is going to beg the question.

[1] See *The Origins of Pragmatism*, part i, chap. 3, section B.

In short, we can prove our procedure to be rational, only if we adopt a standard of rationality which is tailored to our procedure. To this extent Hume is entirely vindicated. But does it then follow that we have no justification for any of our factual inferences? In trying to answer this, I think we should consider what we do in fact regard as an acceptable reason for a belief, beyond a straightforward appeal to perception or to memory. If one proposition is advanced as a ground for another, there are two conditions at least that must be satisfied if the argument is to be acceptable. The evidential proposition has to pass for being true and the connection between the propositions has to pass for being truth-preserving.

I am now going to suggest that these conditions, in a slightly stronger form, can be taken to be not only necessary but sufficient. This leads to the following consequences:

1. We have good reason to accept a singular proposition q on the basis of another singular proposition, or conjunction of singular propositions, p, where p is true, and where the guiding principle of our inference from p to q is a true universal proposition r. Here it is requisite that r should really be used by us as a principle of inference, and that it should not be a generalisation which is constructed *ex post facto*. Otherwise any true singular proposition could be made to furnish a good reason for accepting any other.

2. We have good reason to accept a universal proposition when we deduce it from a true theory or from a more comprehensive generalisation which is true.

3. We have good reason to accept a statistical generalisation when we hold some true theory which authorises our projection of an ascertained frequency, not necessarily to the parent population as a whole, but at least to the further segment of it to which we are interested in extrapolating.

4. In the case where the guiding principle r of our inference from p to q is not a universal proposition, but a proposition of tendency, we have good reason for accepting q, if p and r are true, and if we are not in possession of any other true proposition of tendency, which is stronger than r, and such that it correspondingly favours not-q.

5. In the case where the theory or the universal generalisation, which leads us from p to q, is false, we may still have good reason to accept q if the theory or generalisation entails a strong proposition of tendency which satisfies the conditions laid down in the preceding paragraph. It may also be necessary, here, to put some restriction upon the range

of the proposition of tendency. The purpose of this would be to protect the proposition from being weakened by adverse instances which are remote from us in space or time, and so to avoid our being denied good reason for beliefs which would not play us false in the instances which we actually encountered.

It is to be remarked that what I am requiring in all these cases is just that the premisses and the guiding principles of inference, in terms of which our conclusions are justified, should in fact be true and not that we should know or even have good reason to believe that they are true. And this may well appear unsatisfactory. For what comfort can it be to us, it may be objected, that we have good reasons for our beliefs, if we have no way of knowing that they *are* good reasons? With regard to the premisses, the difficulty is not so serious, since they will be identical with, or inferentially related to, propositions which can be directly justified by perception or by memory. But the generalisations which serve as our guiding principles will at best have been confirmed, and we have not found any means of licensing the passage from the fact that a generalisation has been confirmed, however strongly, to the conclusion that it is true.

I feel the force of this objection. It would be agreeable to have a logic of induction which would underwrite the generalisations on which we rely in arriving at our beliefs about particular matters of fact. And indeed, we can have one, if we stack the cards. We can simply stipulate that it is rational to accept a generalisation when it has acquired a high degree of instance-confirmation, without meeting with any counter-instances, or alternatively, when it has been subjected to severe tests and not been falsified. I do not think that it matters here which way we put it, as we have seen that the difference which is thought to obtain between the approaches represented by Hempel and Popper is not substantial. This comes out clearly when we look at the part that testing plays in Popper's system. For, as I said earlier, the point of testing our hypotheses is surely to accredit them. If all that mattered were their not being found to be violated, we should do better to keep them cloistered.

However, if we do stack the cards, we have to do so carefully. We have to place restrictions, as has been shown, not only on the types of predicates but also on the types of hypothesis that we are going to regard as projectible. It is only by legislation of this kind which takes account of extra-logical considerations, such as simplicity, that we can make it rational to accept any given hypothesis rather than any one of the others that can be devised to accord with our existing information. The point is that our way of looking at the world, as evinced by our conceptual system, our methods of interpreting our observations and our selection of general hypotheses, goes together with our standard of rationality. If someone has an altogether different way of looking at the world, and correspondingly different standard of rationality, we can not prove the superiority of our standpoint except by begging the question. We can only gamble on being more successful and then wait to collect our winnings, or possibly not collect them.

Where I differ from Hume is in not seeing this as a reason for scepticism. The conclusion that all non-contradictory judgements about the future are equally credible is simply false, if we assess it by our own standards. And Hume has not supplied us, and indeed on his own correct principles could not have supplied us, with any logical reason for adopting a standard of rationality which would make it true. In a sense, however, I do not differ from him even at this point. For he ends, as I do, with a superfluous injunction to hold fast to what he calls our natural beliefs – these being in fact the rather sophisticated procedures which we have come to follow in arriving at our beliefs – mainly, if not entirely, because of the success which they have brought us. And even here there is circularity, since we use these procedures in measuring their own success.

II Has Harrod Answered Hume?

Sir Roy Harrod's book on the *Foundations of Inductive Logic* has not attracted the attention which it deserves. A great deal has been written on this subject since 1956, when Harrod's book was published, but I have found very few references to it in the literature and hardly any serious attempt to evaluate its conclusions. Yet there is no doubt that if these conclusions are sound they are, as Harrod claims, of great philosophical importance. For reasons which I shall give, I do not think that he does make his main contention good, but his argument is highly ingenious and his fidelity to the empiricist standpoint, which I share with him, is admirably consistent. Even if his claim to have solved one of the most intractable problems of philosophy is not acceptable, it calls for serious consideration.

Harrod described his book in the preface as the refutation of Hume. Starting from Hume's basic principles, he professes to rebut his sceptical conclusions. He agrees with Hume that inductive reasoning is not demonstrative, and that there is no relation of non-logical necessity by which distinct events could be connected; but he does not agree that there is no valid means of showing that the conclusions of inductive arguments are even probable. He does admit that if we had to make special assumptions about the constitution of nature in order to validate our judgements of probability, inductive reasoning would not be justifiable; and he rightly scorns the device of assigning to empirical propositions an initial probability. Besides the fact that he is not one to accept what Russell has called 'the advantages of theft over honest toil', it is not clear to him, any more than it is to me, what 'initial probability', in this usage, can be understood to mean. He is, therefore, at one with Hume in holding that any judgment of probability which relates to a matter of fact requires to be supported by empirical evidence. But whereas Hume maintained that in cases where two empirical propositions did not stand in any relation of logical entailment it was only in virtue of some factual assumptions that one could count as evidence in favour of the other, Harrod thinks himself able to show that, in certain crucial instances, the relation of 'being evidence in favour of' is a logical relation. His idea is that even when it is granted to Hume that propositions which refer to distinct events are logically independent, it is still permissible for one to entail that another is probable.

The view that probability, in the sense which is here in question, is a logical relation has been held by other writers, including Keynes and Carnap, but Harrod's version of it is original. He ascribes probability to events, rather than to propositions, but his definition can easily be made to apply to propositions. One has only to substitute for the reference to an event a reference to the proposition which states that the event exists. Since it is propositions and not events that are logically related, this would be the more correct formulation; but I prefer to reproduce Harrod's definition in the form in which he gives it.

His first step is to introduce the concept of 'evidential value'. An event is said by him 'to have evidential value if it belongs to a class of events all having a certain character in common, which we may call A. This character A is such that it does not often happen that an event having that character occurs and some other event of a kind determined by the specific nature of the A event does not occur.'[1] What Harrod means by this is not, as one would at first suppose, that there is a high positive correlation between the incidence of the class which is specified by A and that of some other class, but rather that each member of the A class has a co-ordinate in the other class which accompanies it on most occasions. He should, therefore, have made it clear that the events of which he speaks are not individuals but types. An individual event occurs or fails to occur; it is only of a type, or kind, of event that one can significantly say that it occurs in such and such a proportion of cases. At this stage in his exposition Harrod does not refer to these proportions in numerical terms. A little later on, however, he gives it as his opinion that 'observations giving valid grounds for probability judgements always entail a precise number'.[2] If, as very often happens, we are unable to specify the number, it is because we have not been in a position, or not thought it worth while, to make the count. This being his opinion, it is surprising that Harrod should stipulate in his definition of evidential value that it applies only to cases where the positive correlation is high. It would appear more natural to allow that

---

[1] R. F. Harrod, *Foundations of Inductive Logic*, p. 29.   [2] Ibid., p. 35.

any established frequency, however small, has its corresponding evidential value. And this is, in fact, how he usually proceeds.

To say, then, that an individual event *e* is probable to a degree *m/n* is, on Harrod's view, to say that there is some actual event *f* such that events of the kind *F* have the evidential value *m/n* with respect to events of the kind *E*. It is, however, not sufficient, in his opinion, that the *F*s and *E*s should actually occur together with the designated frequency: if the occurrence of an *F* is to be evidence for the occurrence of an *E*, it is necessary also that their connection should be known. But the only way in which it can be known, on the assumption that we have not run through all the instances, or at any rate do not know that we have, is by its having a logical basis. Harrod, therefore, adds the requirement that the *F*s should be of such a nature as to have, by definition, the character *A*, their possession of which logically entails that *E*s accompany them with the frequency in question.

This is a very strong requirement; so strong indeed that it is not at all obvious that anything satisfies it, once we go beyond the purely mathematical calculus of chances into the domain of empirical fact. If it were written into the definition of probability, as Harrod seems to be proposing, the range of true ascriptions of probability would at best be extremely small. In fact, Harrod does not himself adhere to this proposal. He repeatedly speaks of probability in cases where the frequency of the type of event to which the probability is ascribed is certainly not deducible from the evidence with respect to which the probability is assessed. What he requires of these judgements of probability is only that they be justifiable in a roundabout way on the basis of judgements of probability which do satisfy his logical condition.

That Harrod does not, in general, conceive of an ascription of probability as following logically from the evidence with respect to which the probability is assessed is shown by his refusing to adopt the axiom that if the probability that an event *e* will occur, on evidence *f*, is *m/n*, the probability, on the same evidence, that *e* will not occur must be 1 — *m/n*. His reason for this refusal is that it can very easily happen that we have weak evidence in favour of *e*, without its being the case that we have strong evidence, or indeed any evidence at all, in favour of not-*e*. Even if the evidence we have is all in favour of *e*, it may still be weak because it is slender. 'In these circumstances,' he concludes, 'it would be grossly fallacious to argue that there is a high probability that the hypothesis is false.'[1] But while this is a perfectly good argument,

[1] Ibid., p. 34.

the cases to which it applies are just those that do not satisfy Harrod's logical condition. In any case in which the evidence logically entailed that the frequency of the occurrences of events of type E was m/n, it would also entail that the frequency of their non-occurrence was $1 - m/n$. And in fact this applies, as we shall see, to the single example which Harrod offers of a class of cases in which his logical condition is satisfied.

The only axiom which Harrod adopts at this stage is the following: 'If it is the case that B is true if, and only if, A is true, whatever probability pertains to A pertains to B also.'[1] Since he regards this axiom as being of the utmost importance for the development of his system, it is unfortunate that he should have failed to say what meaning he attaches to it. He can hardly have intended his 'if and only if' to be the relation of material implication, since this would have the ridiculous consequence that all true propositions, and likewise all false propositions, had the same probability. A more plausible interpretation would be to take the bi-conditional as stating that A and B were necessary and sufficient for one another, but even so the axiom would be unacceptable. Indeed, the only case in which it would clearly hold would be that in which the proposition that B was necessary and sufficient for A was itself part of the evidence, and there is certainly no reason to suppose that this would always be so. A third possibility is that Harrod was taking A and B to be logically equivalent, and in that case I think that the axiom should be accepted, though even here it leads to difficulties, in the well-known example of 'the infamous ravens',[2] which Harrod does not discuss.

Having given his definition of probability, Harrod then sets out to prove that it is satisfied. He undertakes to show that even on the assumption that 'we are starting from a condition of total nescience',[3] our experience provides us with instances of his character A; that is, a character which is such that from the fact that it is exemplified it follows logically that some other character X, which is not comprised in A, is also exemplified with some determinate frequency. As I said before, it is not at all obvious that any such character is to be found; but Harrod devises a most ingenious example.

What he does is to introduce the notion of a journey which is defined as 'the continuation of a specific uniform feature – colour, pattern or what not'.[4] He then invites us to consider the case of a man

[1] Ibid., p. 48.   [2] See above, pp. 67ff.
[3] Harrod, *Foundations of Inductive Logic*, p. 52.   [4] Ibid., p. 53.

who is embarked on such a journey, in the sense that he is experiencing such a continuity. The man has to know enough to have acquired the concept of the past and future, but he need not be credited with any factual knowledge, beyond the knowledge that he is having the experience in question and has been having it for such and such a length of time. He then forms the hypothesis that the journey will continue for at least $1/x$th of the time that it has already lasted. If he believes this throughout the journey he is bound to be right x times for every one that he is wrong. But this means, according to Harrod, that the probability of his being right on any given occasion is $x/x + 1$. If x is given the value 10, this probability comes to $10/11$. So Harrod is able to cast for his character A the property of being in the course of such a journey and for X the property of having it continue for at least $1/x$th of the time it has already lasted. Alternatively, A could be identified with the property of being a continuity and X with its continuance, or, better still, one could speak of the proposition that there is a continuity of the requisite kind as entailing the proposition that there is such and such a probability that it will continue for such and such a fraction of its existing length.

There is no doubt that this reasoning is mathematically correct. For instance, let us assume that the continuity in question persists for just 11 seconds. Then throughout the first 10 seconds of its duration, the proportion of the time for which it will still endure is always more than $1/10$th of the time for which it has endured already. If it is represented as a line which is divided into eleven segments, it is only in the last segment that the proportion becomes less. Admittedly, this is on the assumption that the line is infinitely divisible. If we believe, as Harrod does, in the existence of *minima sensibilia*, we must recognise that his formula breaks down in the case of any continuity which lasts only so long as a *minimum sensible*; for the question whether the continuity will persist for a fraction of a *minimum sensible* is not significant: there are here only two possibilities; either the continuity will persist for at least as long as it has already lasted, or it will come to an end. Harrod acknowledges this, but fails to see that his formula breaks down also in the case of longer continuities if the length represented by the fraction '$1/x$th of the period already traversed' is less than that of a *minimum sensible*. This places a restriction on the values that can be given to x as well as on the shortness of the continuities, but both these restrictions are slight. They do not seriously impair the generality of Harrod's argument.

Subject, then, to these provisos, we are supposed to be able to infer with certainty, in any case in which we know that a continuity has lasted for a period n, that there is a probability $x/x + 1$ that it will persist for a further period $1/x$th of n. From this two interesting points emerge. The first is that the smaller the extrapolation, the larger the probability. The probability that the continuity will persist for a further $1/10$th of its existing length is $10/11$, the probability that it will persist for a further $1/100$th is $100/101$, for a further $1/1000$th, $1000/1001$, and so on. The second is that since the probability that the continuity will persist for any given fraction of its existing length remains constant; the longer the continuity has lasted, the greater the period that there is a given probability of its persisting. Thus, if it has endured for 10 units of time there is a probability of $10/11$ that it will persist for one further unit, but if it has endured for 100 units, there is the same probability that it will persist for a further 10, if for a 1000, for a further 100, and so on. This is a simple consequence of the mathematical truism that $1/10$th of 10 is 1, $1/10$th of 100 is 10, $1/10$th of 1000 is 100, and in general that the larger a number is, the larger a given fraction of it will be.

That there should be a probability as high as $10/11$ that a continuity which has lasted for a thousand temporal units should persist for a further hundred may seem a very important result. At least it would be important if we were entitled to interpret the probability, in the way that Harrod wishes, as implying that the person who is experiencing the continuity is justified in expecting it to persist for the specified period with the corresponding degree of belief. There are, however, difficulties in the way of this interpretation. To illustrate one of them, let us take up Harrod's simile of the traveller along the line. Ignoring the complication of *minima sensibilia*, we find that the question which he is supposed to be continually asking himself, 'Shall I be able to continue my journey for $1/x$th of the distance that I have already travelled?', receives an affirmative answer in all but the last $1/x + 1$th section of the line. It is from this that the probability $x/x + 1$ is derived. Now the fact that the traveller does not know his position on the line, in the sense that so long as his journey continues he never knows what proportion of its total length he has already traversed, makes no difference to the probability of his receiving an affirmative answer to his question, since wherever he happens to be this probability remains the same. It does, however, make a difference from the point of view of the traveller. Given that his journey is of a certain minimal length, that the value which he gives to x is not too small, and that he asks his question uniformly from the start, the traveller knows *a priori* that the total of affirmative answers to it will exceed the total of negative answers by a calculable amount, but this knowledge is of no interest to him. It tells him only that no matter when the negative answers start coming in their number will be less than that of the affirmative answers which he has already stockpiled. But what he wants to know is when it is likely that the negative answers will start coming in. To put it more precisely, he is interested in the ratio of future affirmative to negative answers from the point which he currently occupies, and to the answer to this question it would seem that nothing in his previous experience of the journey gives him any clue at all. So long as he cannot know what proportion of the total line it represents, the distance which he has already covered becomes irrelevant.

Harrod acknowledges this difficulty, but believes that he has found a way of overcoming it. His method is to represent the total series of answers, that is to say predictions that the journey will continue for at least $1/x$th of the distance already covered, by a straight line of $x + 1$ units. Going from left to right, the x units embody the true answers and the 1 unit those that are false. In principle, each of the units has to be multiplied by a coefficient n, representing the uniform rate at which the answers are being given, but since n cancels out it can be ignored. The next step is to consider the answers from every point of view, that is, from the point of view of every position on the line. This is effected by constructing a square on the original line, as in Fig. 1 (see p. 98).

The horizontal line AB, which now forms the base of the square, continues to represent the series of answers. The vertical line AC represents the number of times this series has to be surveyed if it is surveyed from every point of the journey. Now draw the diagonal AD and let p be a point on AC at a distance y from A. From p draw a line ps parallel to AB and interesecting the diagonal at q. Then pq measures the answers already past and qs those that are future when the traveller has gone the distance y. This may be done for all values of y from 0 to $x + 1$. Consequently, the area to the left of the diagonal represents the answers which are past at the various stages of the journey, and the area to the right of the diagonal those that are future. Now we know that it is only in the last $1/x + 1$th part of the journey that the answers become false. So if we shade the rectangle formed by drawing a vertical line parallel to the sides of the square from the point at which the false answers begin, we can take the shaded area as representing the false answers, and the remainder of the square the true answers. What we now have to calculate is the ratio of true to false answers, given that we take no account of past answers at every stage. This result is yielded by taking the ratio of true to false answers on the right-hand side of the diagonal, and this can easily be seen to be $x^2/(x + 1)^2$.

Here then we have the probability at any stage of the journey that it will be prolonged for at least $1/x$th of the distance already traversed, leaving out of account any stockpiling of answers from the past. It is,

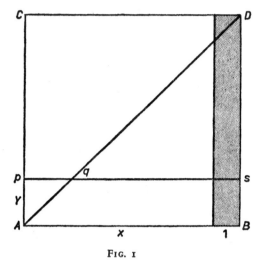

Fig. 1

indeed, a lower probability than our original $x/x + 1$, but it can be made to approach certainty the larger we take x to be, and so the smaller extent of the extrapolation represented by $1/x$. And even though we are not taking any account of past answers, it still remains true that the longer we have travelled the greater is the further distance for which there is a constant probability $x^2/(x + 1)^2$ that we shall be able to continue.

Harrod remarks, correctly, that his formula does not require any assumption about the distribution of continuities in our experience. Neither does it presuppose the principle of indifference in its classical form. The traveller does not need to assume that the line on which he is travelling is as likely to be of any one length as of any other. Whatever its length, provided that it exceeds the requisite minimum, the formula applies equally well. On the other hand, I shall argue that the very fact that the formula holds independently not only of the length of the line but also of the traveller's position on it does introduce the principle of indifference in another guise. How serious this consequence is for Harrod's theory I shall consider later on.

Harrod refers to his conclusion that there is a probability $x^2/(x + 1)^2$ that any continuity will persist for at least $1/x$th of its existing length as the Principle of Experience. It is on this principle that he takes what he calls simple induction to be founded, and it is on the validity of simple induction that all our factual reasoning, in his view, ultimately depends. As we shall see, the principle of experience is indirectly used to justify the ascription of probability to hypotheses which range over the past and present states of the universe; it plays what Harrod calls a primary part in his attempt to justify our trust in memory, and it directly sustains the conjectures that we make about the future.

That Harrod should rely on the principle of experience to justify our trust in memory might be thought to involve him in a circle, since it would appear at first sight that he has to assume the reliability of memory in order to establish the existence of the continuities on which the principle operates. He meets this difficulty by setting himself to show that our trust in memory can be justified merely on the basis of continuities that fall within the specious present. His argument is that in itself the occurrence of any given content of a specious present is very improbable indeed. It is improbable just in the sense that any particular combination of data would be only one out of an enormous number of combinations that were logically possible. On the other hand, if these data were known to be the prolongation of past continuities, they would acquire probability from the principle of experience. But in very many cases, we seem to remember that these past continuities have in fact obtained. We rely on these apparent memories to make predictions that turn out to be successful within the specious present. Since the success of these predictions would be so very improbable if our memory-beliefs were false, there is a high probability that they are true. It is, indeed, only a narrow class of memories that can be vindicated in this manner. We have, however, a way of extending it, if the experiences of which they are memories themselves include predictions which are successful within their specious presents. By this means we can build up a respectable sample of memories which are very likely to be veridical; and then by the use of a principle of fair sampling, which, as will presently be seen, itself is made to depend on the principle of experience, we can secure high probability for a substantial proportion of our memory-beliefs.

I have two comments to make on this argument. The first is that it requires the notion of a continuity to be taken in a much wider sense than that which Harrod originally gave it. If we adhered to the model of a uniform expanse of colour, the only conclusion that we could reach on the basis of the principle of experience would be that the immediate past was exactly like the present. In so far as our apparent memories suggested to us that it was in any way different, the correct inference would be that they were probably delusive. Harrod avoids this unwelcome conclusion, which would make it difficult for him to establish any high probability for the truthfulness of memory in general, by including among his continuities not only the persistence of sensory qualities but also processes of change, and indeed any constant conjunction of disparate events. Thus among his examples of predictions within the specious present that are supposed to supply material for the principle of experience are the prediction that a path will end in a lake and the prediction that when the hands of Harrod's watch point to 9.5 p.m. the bell in Tom Tower will begin to toll. But this is to extend the notion of a continuity so far that it now looks as if any hypothesis which is in tune with past observation can be taken as founding a continuity. Since a 'continuity' is in any case a term of art for Harrod, we cannot object to his using it as he pleases. The difficulty, as we shall see, is just that he may have made it so elastic that it ceases to serve his purpose.

The second point to which I wish to draw attention is that when Harrod speaks of the improbability of our being able to make the successful predictions that we do, if our memories were not veridical, he is not using the term 'probable' in the sense in which he defined it. There is no question here of the existence of any *de facto* correlation between different classes of events, or even of there being such a relation between two different classes of events as logically to entail that their members are correlated with such and such a determinate frequency. The sense of probability which Harrod now brings into play is indeed logical, but it has nothing to do with any actual frequencies. As in the calculus of classes, it relates only to the enumeration of *a priori* possibilities. To say, in this sense, that a state of affairs is probable to the degree $1/n$ is just to say that it is one of n states of affairs which are logically possible. How, in the case of the contents of some present experience, the range of the logical possibilities is to be determined is a question into which Harrod does not enter. The reason why he does not enter into it may be that he had not noticed that his argument requires that it be answered. We have, indeed, already remarked that one of the claims which he makes for his general theory is that it avoids the dubious assumption of initial prior probability. We shall see in a moment that when he engages in the manœuvre of 'reversing the consequences' he replaces this assumption, on which the classical treatment of inverse probability depends, by invoking our knowledge of the ratio of the occurrences of certain kinds of samples. It is, however, obvious that this device will not serve him in the present instance, since we have to rely upon memory at some point in order to determine at what rate samples are occurring. Consequently, the attempt to justify our trust in memory in terms of an estimate of probability which depended on this knowledge would be circular.

If initial prior probabilities are to be admitted, the principle which Harrod's argument requires is that when an event e has a low initial probability but a high probability given some hypothesis h, the occurrence of e bestows a high probability upon h. But now it seems to me that this principle itself is open to question. It involves a tacit passage from the premiss that e is only one of many logically possible alternatives to the conclusion that e is antecedently unlikely to happen; and in the absence of any special assumptions about the constitution of nature, it is not clear why this inference should be thought to be valid, or even what the conclusion is supposed to mean. Neither is it clear what is meant by saying that h is probable, unless this is just a way of issuing us with a licence to accept it. Surely, this is not a probability that can be analysed in terms of any actual frequencies.

Can anything be done to salvage Harrod's justification of the claims of memory, if, in conformity with his declared intention, we refuse to admit initial prior probabilities? The obvious course would be simply to drop the premiss that e is improbable in the absence of h. Then the principle at work would have to be that if an event e, which is not otherwise susceptible of a probability judgement, is highly probable given h, the occurrence of e bestows a high probability upon h. But, apart from the fact that there will still be a problem of interpreting the ascription of probability to h, the condition that e is not otherwise susceptible of a probability judgement is not going to be satisfied. We can imagine any number of past continuities which, in combination with the principle of experience, will yield various probabilities for e. Admittedly, there will be only one set of these continuities of which it is true that we seem to remember that they occurred. But that is no reason to favour them, unless we assume that our seeming to remember something is of evidential value in itself. And this is an assumption which Harrod was trying to avoid.

The main source of these difficulties, it seems to me, is that Harrod is trying to employ what amounts to an argument of inverse probability under conditions where it is not applicable. An argument of this kind depends on our being able to collect samples which may be taken as reflecting the composition of the larger population from which they are drawn. But the contents of a single specious present do not provide any such samples: in the state of total nescience, in which Harrod places us, we are not even entitled to assume that the larger population exists. Even if we were to take the reliability of memory for granted, we should hardly be able to compile from this source alone a stock of samples from which we could extrapolate with any degree of confidence. One needs the further evidence which is obtained by drawing on the testimony of others and on historical records. This illustrates the general difficulty which besets those who look to the theory of probability to justify induction. They have to use inductive processes in gathering the material on which the theory of probability can be put to work. Harrod is more keenly aware than most of the threat of this circle, but I do not think that he entirely succeeds in avoiding it. I do not see, indeed, how it can be avoided.

Let us, however, set this difficulty aside. Let it be assumed that we have a respectable amount of trustworthy information about the results of observations which have been made up to the present time. We can then look to arguments of inverse probability to supply us with probable conclusions about the general characteristics of the things on which these observations bear. Harrod confines these conclusions to the period of time which the observations cover, leaving it to the principle of experience to buttress any conjectures that we make about the future. There is, however, no reason in logic why the scope of the conclusions should be restricted in this way. The reason which Harrod gives, that the future is outside our range of observation, is not cogent, since our range of observation is limited, in any given instance, by the position which we occupy in space as well as by the position which we occupy in time. The point of resorting to inverse probability is that it carries us beyond these limits; and this applies to an extension in time no less than to an extension in space.

The principle on which arguments of inverse probability depend is that of the Law of Large Numbers.[1] Though the operation of this law is quite general, Harrod chooses to concentrate exclusively on cases in which the character for which he is sampling occurs in the sample with a frequency of 100 per cent. His reason for this is that owing, as he puts it, to the fact that 'the frequency curve has a sharp declivity when it approaches all or none'[2] he is able in this way to obtain higher probabilities on the basis of smaller samples. Thus he calculates that, given a population which may be as large as one pleases, in which 95 per cent of Ps have the property Q, a probability that in a sample of 540 members all the Ps have Q is 1/1,080,000,000,000. That is to say, there are over a billion possible samples in which at least one P lacks Q for every one in which all Ps have Q. Harrod characterises this fact by saying that in relation to the hypothesis that only 95 per cent of Ps have Q a sample of 540 in which all Ps have Q is a billion-fold deceptive, and accordingly that it is a billion-fold suggestive in relation to the hypothesis that the proportion of Ps which have Q in the total population of Ps is higher than 95 per cent.

Does it follow that we have good reason to believe that the proportion of Ps which have Q in the total population of Ps really is higher than 95 per cent? What we know is that if it is not higher, our sample is very untypical, in the sense that it belongs to a very small minority, not necessarily of actually recorded samples, but of all possible samples of its own size. But why should it not be untypical in this sense? What assurance do we have that our actual samples are not deviant?

Harrod meets this difficulty by introducing a postulate of fair sampling. He does so on the assumption that it can later be shown to be unnecessary, so that the part which it plays in his system is that of what he calls a tool of thought. His postulate is not the simple but implausible one that any given sample is fair but rather, as he puts it, that 'on a long run of experience our sample of samples is fair'.[3]

This assures us that 'If we are confronted with many populations in which 95 per cent of Ps have Q, we shall only come across a sample of 540 in which all Ps have Q once for every billion times that we come across a sample of 540 in which at least one P lacks Q'.[4] Now suppose that we are able to ascertain the rate at which such billion-fold suggestive samples are in fact occurring in our experience. For the sake of the argument, let us take this rate to be once per 1000 samples. Then

[1] See above, pp. 40–1. [2] Harrod, *Foundations of Inductive Logic*, p. 105.
[3] Ibid., p. 89. [4] Ibid., p. 90.

the fair-sampling postulate guarantees that 'If we believe a hypothesis supported by evidence of this kind we shall be right 1000 million times for every once that we are wrong'.[1]

The reference to many populations and to our being right a given proportion of times shows that Harrod's postulate is more complex than might at first appear. He is not assuming that if we have a sufficiently numerous set of samples which are drawn from a given population, the average distribution of the character throughout the samples approximately matches its distribution in the population as a whole. His samples do not all relate to the same hypothesis: if they did, they would be drawn from one population, not from many, and there would be no question of our being sometimes right and sometimes wrong. If the hypothesis in question were true, we should be right all the time, and if it were false we should be wrong all the time. The samples, therefore, relate to different hypotheses, and what the fair-sampling principle is meant to ensure is that the rate at which 100 per cent samples of whatever kind occur throughout the whole of our experience corresponds to the distribution of the logical possibilities. What Harrod is in fact doing is to treat 100 per cent samples which conform to different hypotheses as if they conformed to the same hypothesis, and then apply the law of large numbers.

The advantage of this procedure is that it supplies us with a more numerous set of samples than we could in practice obtain by confining our selections to a single class of objects, but there are serious objections to it. Harrod himself admits that it appears paradoxical that 'the evidential value of a specific piece of evidence for a specific hypothesis depends on the rate of occurrence of pieces of evidence that have no specific bearing on the hypothesis at all',[2] but argues that this sense of paradox should not survive closer scrutiny. If part of what we mean by saying that a conclusion is probable is that 'it does not often happen *both* to have evidence of the kind that is before us *and* for the conclusion to which it points not to be the case',[3] we are in any case referring to evidence which is not specifically of the kind that is before us but only of the same logical structure. In fact, this does not follow. We should satisfy the definition if we argued from the fact that the last *n* *A*s we had observed were *B*, to the probability that the next *A* to come to our notice would be *B*, without taking account of any other sort of evidence than the frequency with which a series of *n* *AB*s had been followed by an *n* + 1th. What is true is that if we ascribe probability

---

[1] Ibid., p. 93.   [2] Ibid., p. 100.   [3] Ibid.

in this sense to a universal hypothesis, we must be taking account of a wider range of evidence, for the reason, which I have already mentioned, that it makes no sense to speak of a specific hypothesis of this kind as being often or seldom the case; it is only when we view it as a member of a class of hypotheses that the notion of truth-frequency becomes applicable to it. This might be taken as vindicating Harrod's approach, but it also might be taken as showing that his definition broke down when it came to the probability of universal hypotheses: that it properly applied only to conclusions which referred to types of particular events.

I myself should be inclined to take the second course, but I do not want to press the argument here. My main objection to Harrod's procedure is that, in the absence of any ruling as to what samples are admissible, its working is almost entirely arbitrary. It is clear from his talk of Ps and Qs that what Harrod means by a sample is a set of individuals which have been selected on the basis of their having some character, or set of characters, P, in common: the question then is whether they also have in common some other character Q. But now if no restriction is going to be placed upon the choice of Q, it is clear that this question can always be answered in the affirmative. Even if we make the stipulation, which Harrod presumably intends, that Q should be logically independent of P, we can still make practically sure of getting a positive answer by fixing on some pervasive property, like that of being more than ten miles from the fixed stars. It can then be contrived that 100 per cent samples occur at the rate of 100 per cent, with the result that all our judgements of probability of the sort in question will, by Harrod's procedure, be ridiculously inflated. Of course there will still be any number of properties the distribution of which throughout a given sample is found to be less than 100 per cent, but if we are going to require for anything to be a 100 per cent sample that its constituents have all their properties in common, we pass to the opposite extreme: the rate at which 100 per cent samples occur in our experience will then be 0 per cent, and in the resulting deflation none of our so-called suggestive samples will count for anything at all. Clearly what Harrod has in mind is something between these two extremes, but he does not say how it is to be arrived at. One possibility would be to confine ourselves to what Peirce called 'pre-designated' characters. This would mean that the only samples of which we took account would be those that were related to the relative incidence of characters which we had already decided to investigate. We could still artificially increase probabilities by choosing to investigate characters which were obviously pervasive, but it could perhaps be assumed that we should not do this, that it would not be regarded as playing the game. It would, indeed, appear from other things that Harrod says that he would consider this limitation to be too restrictive, but it is not easy to see what he could substitute for it. Until this point is decided, the value of this principle of fair sampling must remain in doubt.

This is not, however, a very serious consequence for Harrod, if he is able to show that his principle of fair sampling plays no greater role in his system than that of a dispensable tool of thought. The argument which he offers for its being dispensable is very simple. Either the apparent stability and uniformity which our observations have revealed to us reflect the constitution of the universe, or there has been a very great systematic bias in our sampling. If there has been a bias we can infer by the principle of experience that it is likely to continue. Accordingly, whether our sample of samples is fair or not, the extrapolation of our evidence to further instances is likely to be successful; and this is the only assurance that we need.

A weakness in this argument is that it assumes that we have a way of determining the degree of regularity that has so far been displayed by the phenomena which have come under our observation. To the extent that this remains arbitrary, as I have argued that it does under Harrod's treatment, it will be correspondingly unclear what future expectations we are entitled to cherish. It would therefore seem preferable to apply the principle of experience, not to anything so general and vague as the apparent order of nature, but directly to the specific hypotheses that we want to project. This also has the advantage that we are spared the dubious step of treating the apparent order of nature as being yet another sort of continuity.

Everything, then, depends on the principle of experience. We have seen that the mathematical certainty on which this principle rests is that if a quantity y is divisible into $x + 1$ equal units, then for any number n which is less than y but greater than 1, $y - [(y - n)/x]$ is greater than $y - [(y - 1)/x]$. In the special case in which y represents the length of a line, this simple truism has the consequence that if the line is traversed from left to right it is only in the last $1/x + 1$th part of the line that the remainder becomes less than $1/x$th of the distance already covered. We have, however, also seen that even if we allow this fact to be expressed by saying that, for a traveller along the line, there is a constant probability $x/x + 1$ that his journey will continue for $1/x$th of its existing length, we cannot take this probability to be the measure of the traveller's rational expectation. What concerns the traveller is the proportion of future positive answers, whereas, on the assumption that the question is continuously asked, the formula $x/x + 1$ gives the proportion of all positive answers, past and future. It was to avoid this objection that Harrod substituted the formula $x^2/(x + 1)^2$ which is obtained, as we have seen, by a process of discounting past successes.

Does this second formula meet the difficulty? I do not think that it does. It does not discount the past in the way that is required. What the formula yields is the result of neglecting past successes from the point of view of every position on the line; what it does not yield is the result of neglecting past successes from any given point of view. But this makes it useless to the traveller. It is of no interest to him to know that the ratio of future successes to failures in general is $x^2/(x + 1)^2$. He wants to know what the proportion of his future successes will be, from the position which he actually occupies, and this the formula does not tell him. Since it treats his position as an unknown factor, it is not to be expected that it should.

It would appear, then, that the principle of indifference, in the form of the assumption that the traveller is as likely on any given occasion to be at any one point in the line as at any other, is required after all. Harrod dismisses this as the triviality that the traveller spends an equal period of time in every equal portion of the line; which is, indeed, a triviality, if we think of the traveller as experiencing a temporal continuity. But here he underestimates the difficulty. Let us suppose that the question, in concrete terms, is whether a specified conjunction of properties, which has hitherto obtained in our experience, will hold good in the future. The proposition on which Harrod relies is that, even if there is a counter-example to the hypothesis that the properties in question are universally conjoined, it is unlikely to occur in the immediate future. In the terms of his simile, it is unlikely that the traveller is near the end of the line. But what is the ground for this proposition? Certainly not the simple tautology that equal periods of time are equal. Its only basis is the fact that positions near the end of the line amount to only a small fraction of the number of positions that the traveller could occupy. But clearly this is insufficient, unless we assume that his occupation of any of these positions is antecedently equi-probable; and this is just the sort of assumption that Harrod was trying to avoid. We again find him in the situation of tacitly depending on a notion to which he officially, and in my opinion rightly, denies any meaning.

A postulate which would meet the case is that which Harrod himself puts forward as justifying what he calls conditional simple induction. In the case of unconditional simple induction, it is assumed that the answers to the question whether the continuity will persist for at least $1/x$th of its existing length are given continuously throughout: the induction is said to be conditional when this assumption is dropped. The postulate, which is made necessary by the admission that the question may be raised sporadically, is that 'If a man makes a great many enquiries concerning the continuance of continuities, these will be equi-proportionally dispersed among the sectors of the various continuities'.[1] Harrod says of this postulate that 'it is not a postulate about nature; it is concerned only with the distribution of enquiries about nature',[2] but this distinction must appear tenuous when we see that to ask when the enquiry is made is equivalent to asking at what point in the continuity the traveller makes it, and that this is equivalent to asking how long the continuity has still to run. In effect, Harrod's postulate assigns equi-probability to the traveller's being in any equal section of any given continuity; and this entails the requisite conclusion that he is unlikely to be near the end.

Harrod describes this consequence of his postulate as a weak form of it and claims that it can be justified on the ground that our experience supports it. 'We have not found, as a matter of plain fact, that no sooner do we notice some continuity than it presently comes to an end.'[3] He realises that this argument itself rests on the inductive assumption that the 'fact' in question can be extrapolated to the future, but he thinks that he can meet the difficulty by treating his postulate in its weak form as a continuity and then appealing to unconditional simple induction. 'By unconditional simple induction', he says, 'we are not likely to be on the extreme edge of such an experience.'[4] But for this argument to succeed, it is essential that unconditional simple induction should do the work required of it, without the assistance of any such postulate. If I am right in thinking that this is not possible, Harrod's argument fails.

Finally, I think it has to be said that even if the objections which I have brought against the principle of experience could be overcome, its yield would be much smaller than might at first appear. The reason for

[1] Ibid., p. 65.   [2] Ibid., p. 66.
[3] Ibid., p. 7.   [4] Ibid., p. 77.

this is that we should still be left with Goodman's difficulty[1] that if we put no restriction on the form of our hypotheses, we shall find that any evidence which confirms a given hypothesis *h* will equally confirm some other hypothesis *h'* which entails not-*h*. In Harrod's terms, any path that we traverse will be common to a number of different possible continuities; and any justification that we may have for projecting it in one way will be equally a justification for projecting it in another. This can be simply illustrated by the example of drawing balls from a bag. If there are a hundred balls in the bag and ninety-nine of them have been drawn and found to be blue, we may think that the hypothesis that all of them are blue has been very strongly confirmed. The fact is, however, that in default of any other information, we have no more reason to expect the hundredth ball to be blue than we have to expect it to be any other colour. It might be argued that only if the hundredth ball were blue would the sequence form a continuity, but even to say this would be arbitrary. In an experiment in which we went through a series of bags, a sequence in which one ball in every bag was of a different colour from the others would be just as much a continuity as one in which each bag contained only balls of the same colour.

It might be thought that Harrod had dealt with this objection in his attempt to justify our preference for simple laws. Following Professor Jeffreys, he assumes that laws can be ranked in order of complexity, possibly in terms of 'the number of adjustable parameters' which are required for the expression of the laws. Then he argues that we are right to prefer simple laws on the ground that the odds against getting a set of observations which conform to a simple law are greater than those against getting a set of observations which conform to a more complex law; from which it follows, by inverse probability, that the observations favouring the simple law have greater evidential value. Accordingly, he advances, as a first principle of induction, the proposition that: 'A set of observations conforming to a law has a cogency in establishing the truth of that law that is in inverse ratio to the number of sets of observations which would occur in the absence of any law and conform to *any law of equal simplicity ranking* with that to which the observed set conforms.'[2]

This is an ingenious suggestion; but once again it depends on the forbidden assumption of prior probabilities. Simple laws are supposed to derive their greater probability from the mathematical fact that out of all the sets of *n* observations that could possibly be made, the number of

[1] See above, pp. 82 ff.   [2] Harrod, *Foundations of Inductive Logic*, p. 185.

sets that conform to a simple law is smaller than the number of those that conform to a more complex law. In other words, sets which conform to a simple law are comparatively rare, and their comparative rarity increases with the increase of n. But, quite apart from the difficulty of specifying the totality of possible observations, this mathematical fact supplies no basis for a judgement of probability unless it is assumed that all these observations are antecedently equi-probable, in the sense that each has an equal chance of being made; and again this is a proposition to which, on his own correct principles, Harrod is unable to attach any meaning.

I conclude that Harrod has not found the answer to Hume. He has made a most valiant attempt to justify induction on no other basis than that of the necessary propositions of formal logic and pure mathematics, but his materials are insufficient for the task. If his theory fails, I think that the moral to be drawn is that no theory of this logical type is going to be successful.

III The Problem of Conditionals

A. THE BASIS OF FACT

Conditional statements play a much greater part in our discourse than appears at first sight. To begin with, many of the predicates that we commonly employ are dispositional. They are used to refer not, or not only, to what is actually to be found at such and such points in space and time but to what would be there to be found if the appropriate conditions were realised. This is obviously true of predicates like 'soluble' or 'brittle' or 'irascible', but it is also true of a great many predicates which pass for being purely descriptive, including many of those, like 'being a chair' or 'being a pen' or 'being a match', which are used to identify objects as being of certain kinds. For it enters into the definition of these kinds that the objects which belong to them should have certain causal properties, and the ascription of causal properties is hypothetical: it relates not to the way in which an object actually does behave – a particular match may never in fact be struck, a particular pen may never in fact be used to write with – but to the way in which it would behave in the relevant circumstances. Another way of putting this would be to say that to the extent that objects of different kinds are distinguished by their powers, the statements which identify objects as being of such kinds are implicitly law-like; and it is a characteristic of a law-like statement that it does not so much describe what actually happens as determine what is possible. How far this is a satisfactory method of differentiating generalisations of law from generalisations of fact is a question on which I shall touch later on.

The difficulty, indeed, is to find statements which are purely categorical. The most promising candidates are those in which we refer to an object by a demonstrative, or by the use of spatio-temporal co-ordinates, and attribute to it some ostensive predicate, as, for example, a predicate descriptive of its shape or size or colour. Even so, it is arguable that to predicate a particular size or shape, at least of a physical object, is to refer to the hypothetical results of certain processes of measurement, and that even the attribution of a colour is conditional, in that it refers not to any appearance which the object actually presents on some particular occasion, but rather to the appearance which it standardly would present in normal circumstances to a normal observer.

At this point we have to take a decision. Except at the level of sense-qualia, where it is a question only of what is sensibly presented at a given instant to what is in fact, though not at this level said to be, a particular observer, any statement which describes a sufficient test for the presence of even an ostensive property is bound to be conditional. Apart from any other consideration, this follows from the fact that the property will not be held to be genuinely present unless it is detectable by any normal observer; and the same of course applies to the physical object to which the property is attributed. Consequently, if such attributive statements are construed as implying that the relevant tests are satisfied by the objects and the properties in question, these statements also will have to be regarded as conditional. I think, however, that it is open to us not to construe them in this way. We can, perhaps, detach the content of a predicate from the criteria which we use to determine whether it is satisfied: so that a statement which assigns a physical location to an observable property can be understood realistically, as describing what is to be found at the designated place and time, without containing any reference to the persons who might find it. In short, we can choose to treat such statements as being categorical.

I repeat that it is a matter of choice. To the extent that we can distinguish different levels of theory within our whole corpus of beliefs, the statements at the higher theoretical level are always conditional with respect to the lower-level statements which support them. Thus, the statements of physical theory are conditional with respect to statements relating to physical objects at the level of common sense; and these in their turn are conditional with respect to the experiential statements which monitor sense-qualia. If we are concerned with the theory of knowledge, I believe that we have to take experiential statements as primitive, but, as I have argued elsewhere,[1] we are free to take a different course when it comes to saying what we think there is. The reason for this, in my view, is that it is only at some level of theory that we can form any picture of an objective world. At the very least, it seems to me, we must make such projections of sense-qualia as will be sufficient to yield a spatio-temporal system with what, on the

[1] See *The Origins of Pragmatism*, part ii, chap. 3, section c, and *Russell and Moore: The Analytical Heritage*, chap. 3.

model of Russell's world of sensibilia, I call standardised percepts for its constituents. I propose, then, to look upon this world as constituting a bedrock of fact, and the only statements which I shall regard as being strictly factual will be those that are limited in their content to supplying true or false descriptions of this world, together with such statements as are obtainable from them by quantification or by the use of extensional operators. All other empirical statements, or at least all those that function at a higher level, will be construed as relating to the arrangement, or the explanation, of what are taken to be the primary facts. The distinction which I am here making between facts and their arrangement is due to C. S. Peirce[1] and it corresponds roughly to the distinction which F. P. Ramsey drew between a primary system of facts and a secondary system of theory.[2]

The Russellian world which constitutes my primary system is not quite the world of common sense. Or rather, to put it more accurately, the words which we use to refer to the physical objects of common sense are not entirely suited to describing the objects of my primary system. The reason why they are not is that they are overcharged with references to causal properties. This is not to say that causal statements do not apply to the objects of the primary system, but only that they are not, in my sense, strictly factual. They do, indeed, have a factual content but they overflow into a secondary system, which is concerned with the arrangement of facts or what are taken to be so. In short, I want to make a distinction which ordinary language blurs. Thus the familiar objects which will figure in my examples are meant to be identified only by their phenomenal properties. My references to them are not to be understood as carrying any logical implications about their powers.

B. NON-TRUTH-FUNCTIONAL CONDITIONALS

So long as we keep within the domain of fact, as I have just defined it, the only form of conditional which will be at our disposal is that which is known as the material conditional. In other words, we shall have to regard the statement-form 'If p then q' as being equivalent to 'not-p or q'. It is, however, notorious that very few of the conditionals that we

[1] See my *The Origins of Pragmatism*, part i, chap. 2, section B.
[2] See *The Foundations of Mathematics* (1931), pp. 212-36.

actually employ fit naturally into this mould. For example, if I know that a motion was passed at some meeting unanimously with no abstentions, I am prepared to assert of anyone that if he was present at the meeting, he voted for the motion, meaning by this no more than that either he voted for the motion or he was not present. It does not follow, however, that I am prepared to accept it as true of everyone who was not present that if he had been present he would have voted for the motion, irrespective of his opinions on the issue. If there is anyone of whom I am prepared to assert that had he been present he would have voted against the motion, I regard this as incompatible with the assertion that he would have voted for it, although if I interpreted them as material conditionals I should have to say that, so far from being incompatible, both assertions were true. This shows that we do not regard it as sufficient for the truth of such conditionals that their antecedents are false. And the great majority of the conditionals that we think it worth putting forward belong to this second class. In particular, it includes those that are concealed in the use of dispositional statements. The mere fact that this table is never immersed in water is not taken as entailing that it is soluble.

Not only are conditionals of this most common type not thought to be rendered true by the falsity of their antecedents, but it is doubtful if even the truth of both their antecedents and their consequents would always be considered sufficient for their truth. For instance, the man in Sam Weller's story in *The Pickwick Papers*, who when told by his doctor that if he went on eating such a large quantity of crumpets he would be dead within six months, responded by eating an even larger quantity of crumpets and then shooting himself, was trying to refute the doctor, not to prove him right. He took the doctor to be asserting that he would die in consequence of over-eating and therefore set out to falsify his conditional, actually by making both its antecedent and its consequent true. He thought it would be enough if he broke the implied connection between them. Whether he was right may be disputed, since the doctor's conditional could be interpreted as carrying an implied proviso that other things should remain equal. But at least it would seem incorrect to say that the conditional had been verified.

Because the examples usually chosen to illustrate my second class of conditionals are examples in which the antecedent is taken to be false, it has become a common practice to refer to these conditionals as counterfactuals. There are, however, reasons why this is not a good classification. In the first place, as our example has shown, there can be material conditionals with false antecedents. If in fact the voting was unanimous, then to say of someone that if he was present he voted for the motion is equally true whether he was present or not. In the second place, as we have also seen, a conditional may have a true antecedent even though it is not material. This is shown by the example of the crumpet-eater. And finally, to talk of counterfactuals is to draw attention to the wrong point. What makes these conditionals interesting is not the truth or falsity of their antecedents but the fact that they are not truth-functional. Their truth-value is determined by the truth-value of their components only to the extent that the truth of their antecedents combined with the falsity of their consequents is sufficient, though arguably not necessary, to make them false; but, as we have seen, neither the falsity of their antecedents, nor the truth of both their antecedents and consequents, is sufficient to make them true.

A slightly better way of classifying these conditionals, which is also current, is to call them subjunctive, but this, too, is not entirely accurate. No doubt all the conditionals which we should naturally express by using the subjunctive mood are non-truth-functional, but the converse does not hold, as the example of the crumpet-eater shows once again. The difference between saying 'If you continue to eat so many crumpets you will be dead in six months' and saying 'If you continued to eat so many crumpets, you would be dead in six months' is that in the second case some doubt is conversationally implied as to whether you will continue, perhaps even a slight presumption that you will not. In the case of a past subjunctive, as in our example 'If he had been present, he would have voted for the motion', the presumption is stronger, to the point that it can be construed as logically implying that the antecedent is false. But if the interest of such a conditional does not consist in its antecedent's happening to be false, it equally does not consist in any implication or suggestion that the antecedent is false. In calling these conditionals subjunctive, one is, therefore, again drawing attention to the wrong point. This being so, it will be better, perhaps, to sacrifice elegance and speak of them simply as non-truth-functional.

The first question that arises is whether we can give an adequate account of these non-truth-functional conditionals within the boundaries of what I have called the domain of fact. Since the language which I am taking to be descriptive of fact contains only extensional operators, it might be thought that this question had already been answered negatively, but this conclusion would be premature. It is, indeed, clear that these conditionals can not themselves be factual statements, in my restricted sense of this term, but it is still arguable that they can be analysed in terms of factual statements and of the logical relations which obtain between them. On this view, non-truth-functional conditionals are statements of the second order. They may comprise factual statements, but they are also used to talk about them.

C. THE META-LINGUISTIC APPROACH

This meta-linguistic treatment of non-truth-functional conditionals has been most fully developed by Professor Nelson Goodman. His investigations, as set out in his book *Fact, Fiction and Forecast*, make it clear that the difficulties in the way of such an approach are very much greater than might at first be thought. He concerns himself only with counterfactuals in the strongest sense, in which both the antecedent and the consequent of the conditional are false, but his results apply more widely. His aim being to lay down the conditions under which a counterfactual of this sort is true, he eventually arrives at the following rule: A counterfactual 'if A then C' is true if and only if there is some set S of true sentences such that S is compatible with C and with not-C and does not follow by law from A, and such that the conjunction of A and S is self-compatible and leads by law to C; while there is no set of true sentences S' such that S' is compatible with C and with not-C and does not follow by law from A, and such that the conjunction of A and S' is self-compatible and leads by law to not-C. Goodman has shown that no simpler formula will work, but now discovers that even this one is too simple. As he himself puts it, 'the requirement that A.S be self-compatible is not strong enough: for S might comprise true sentences that although compatible with A, were such that they would not be true if A were true'.[1] Thus, on the assumption that we have among true sentences a set of sentences referring to the relevant occasion and descriptive of the conditions under which matches ignite, our formula will allow us to ascribe truth to the conditional that if match *m* had been struck it would have lit; but equally, on the basis of the fact that oxygen was present, that match *m* was well made and so forth, together with the true sentence that match *m* did not light, it allows us to derive the false conditional that if match *m* had been struck it would not have been dry. The trouble is that while the

[1] N. Goodman, *Fact, Fiction and Forecast*, 2nd ed., pp. 14–15.

statements that the match was struck and that it did not light are logically compatible they are not, in the circumstances, causally compatible: had the match in fact been struck, it would not have been true that it did not light. Goodman expresses this by saying that the two sentences are not co-tenable. But now if we strengthen our formula by requiring that the conjunction of A and S be not merely compatible but co-tenable, we run into a vicious regress. For to say that A and S are co-tenable is to say that it is not the case that if A were true S would not be true, and to determine whether this counterfactual is true or not we have to enquire whether there is a set of true sentences S'' such that in combination with its antecedent they lead by law to not-S, and such that they and the antecedent are co-tenable. In short, to establish co-tenability in one place, we have to assume it in another.

Another difficulty is evinced in Goodman's use of the expression 'leads by law to'. It might be thought that the appeal to the undefined notion of natural law could be avoided by including the appropriate generalisation in the set of true sentences S and then requiring that A and S entail C. Thus, in the example of the match, one would add to the set of statements which record the fulfilment of the appropriate initial conditions the factual generalisation that under these conditions all matches light when they are struck. But the trouble with this is that not every true generalisation is thought to license a non-truth-functional conditional. If I conjoin the statement that Smith was at the meeting with the true generalisation that everyone who was at the meeting voted for the motion, I can derive the conclusion that Smith voted for the motion, but this does not entitle me to assert that if Smith had been at the meeting, he would have voted for the motion. The reason why it does not is that the true generalisation on which I am relying is not a generalisation of law. But then how do we decide what is a generalisation of law? How does it differ from a generalisation of fact? We can not be content to say that it is the mark of a generalisation of law that it entails non-truth-functional conditionals, for we are trying to explain such conditionals in terms of the notion of law. All that emerges is that the two problems go together.

So, as Goodman admits, his attempt to construe counter-factuals as second-order statements runs into an impasse. And indeed, it is in any case rather doubtful whether an approach of this kind can yield a correct analysis of non-truth-functional conditionals. It may well be that I should not venture to make any such statement as that if Hannibal had besieged Rome after the battle of Cannae he would have taken it, unless I were in a position to support it with some such set of statements as that the morale of the Carthaginian army was high, and that of the Roman army low, that Hannibal was a more gifted general than any of his Roman adversaries, that he had sufficient siege equipment and so forth. Even so, it is not altogether plausible to say that when I put forward the conditional I am implicitly making all these statements, and it is still less plausible to say that I am also making the meta-linguistic statement that they in conjunction with the false statement that Hannibal besieged Rome and some true generalisation of law entail the false statement that Hannibal took Rome. For one thing, I should have difficulty in formulating the generalisation. It would have to be some-thing to the effect that whenever highly gifted generals with victorious armies besieged cities garrisoned by demoralised troops they capture them. But this is hardly a law. I am not sure that it is even invariably true.

The fact is that if we are looking for a general account of the be-haviour of non-truth-functional conditionals, or even of the sub-class of counterfactuals, the requirement that the antecedent should be connected with the consequent by a generalisation of law is too strong. We often commit ourselves to a counterfactual on the basis of nothing more than a prevalent tendency. 'If he had come to dinner, he would have told his story about meeting the Duke of Edinburgh.' It is not a law of nature that he tells this story whenever he goes out to dinner. He does not even tell it on all occasions; but on this occasion I think he would have. Here my counterfactual does have statistical backing, but I am surely not asserting that its consequent is derivable from its ante-cedent with the help of a proportional syllogism. In any case, the derivation would be invalid.[1]

D. TRUTH AND ACCEPTABILITY

It is not clear even that one can not meaningfully express a non-truth-functional conditional without the support of any generalisation at all, however weak. I make a bet: 'If I toss this penny it will come up heads'; and then I toss it and am proved right or wrong. But now suppose that I find no takers, and so do not toss the penny, I then say: 'You were quite right not to bet with me, for if I had tossed the penny it would

[1] See above, p. 51.

have come up heads.' I have no grounds for this assertion – the penny is not two-headed or known to be biased – but I make it all the same. What view, then, should we take of it? Should we say that it is meaningless, because it has no backing? Or that it has no truth-value? Or that it has a truth-value, but one which we can not discover?

It does not seem right to say that my statement would be meaningless. Suppose that I had made my bet in the form: if I were to toss this penny, it would come up heads. It would be strange if my statement could not be known to have any meaning, until it was known whether I tossed the penny or not. And if this conditional can be allowed to have a meaning independently of its antecedent's being satisfied, there appears to be no good reason why the same should not apply when the tense of the subjunctive is shifted to the past.

There is a much stronger case for saying that my conditional has no truth-value when its antecedent is not satisfied. The argument would be that if it is not to be treated as a material conditional, the only basis there can be for predicating either truth or falsehood of it is that we have evidence which points one way or the other; and in the present examples this basis is lacking. There would, indeed, be no very serious objection to treating it as a material conditional. The question is whether we mind saying that it is true both that the penny would have come up heads and that it would not, just so long as it was not tossed.

But why should this be the only other possibility? Why should we not say that there is a truth here which we can not discover? Either the penny would have come up heads or it would have come up tails, but unfortunately we shall never know which. The answer is that it is, indeed, open to anyone to take this realistic view. If he does, he can not be refuted. The most we can say is that his statement is idle. When he employs the concept of truth in this way, he is, in Wittgenstein's simile, letting the machine run, without putting it to work.

But is his statement any more idle than some of those which we are admitting as statements of fact? We are allowing the principle of excluded middle to apply to such statements as that there are flowers growing in some unexplored cranny of a mountain, or shoals of fish swimming in a sea, even though their presence goes for ever undetected. In these cases we can, indeed, claim that the statements are testable in principle. A suitably equipped observer might have occupied the requisite spatio-temporal position. But then I might have tossed the penny. What is the difference?

The answer is that we draw a distinction between the cases in which the lack of verification is due to the absence of an observer and those in which it is due to the omission of some action from which one or other state of affairs would result. Statements relating to the first class of cases are, in my sense, factual; those relating to the second class are not. The underlying assumption is that the processes of observation are not actions which affect what is observed. Or rather, to put it more accurately, we set the level of fact at a point where their contribution has already been taken into account. The distinction is tenuous and, as I said earlier, to some degree arbitrary, but nevertheless one that I think it important to maintain.

Accordingly, I want to say that the only factual assertion that I am making when I say that if the penny were tossed it would come up heads is one which is expressible by the material conditional which states that either the penny comes up heads at the time in question or it is not tossed. Beyond that I am expressing a willingness to bet on its coming up heads, on the supposition of its being tossed. The analogy of betting fits less well when the conditional takes the form of a past subjunctive, but the force of my assertion remains the same. Though speaking at a later time, I put myself imaginatively at an earlier time than that of the hypothetical event. If what I have called the expression of a willingness to bet is construed as a statement, it is one that has no truth-value.

Let us now modify the example by supposing that I have reason to believe that the penny is biased to the extent that it comes up heads three times out of four. What difference does this make to the interpretation of my statement? So far as its meaning and its factual content are concerned, I want to say that it makes no difference at all. The difference is that my statement now has a point which it did not have before. If I am asked 'Why do you think that the penny would have come up heads?' I can give a reasonable answer. It is, however, in no way a decisive answer. If someone says: 'Granted that this penny mostly does come up heads, there are times when it doesn't and why should not this have been one of them?', I have nothing better to reply than 'Why should it have been?'

Now that my statement has a point, does it also acquire a truth-value independently of the truth of the material conditional? I am inclined to say not. It has, indeed, acquired some title to acceptance and we do, indeed, predicate truth just of those statements that we accept. On the other hand, we also have the idea that a true statement should be one that corresponds to a fact; and my statement does not correspond to any fact. Even if it be granted that the frequency with which the penny has been found to come up heads can legitimately be projected on to the results of future tosses, nothing follows with respect to any occasion on which it is not actually tossed.

The same consideration applies, in my view, to the case in which the non-truth-functional conditional is sustained by a universal law. Let it be assumed that we can specify a set of conditions, including the fact that oxygen is present, that the match is dry and so forth, which are such that whenever under these conditions a match is struck it lights. And suppose now, believing these conditions to obtain, I take out a match and say that if I were to strike it, it would light. What factual statement am I making, over and above the material conditional that the match will either light or not be struck?

One obvious suggestion is that I am asserting the existence of the conditions in question; or rather, since many of the statements in which these conditions are set out will themselves be covertly dispositional, that I am asserting the presence of the facts on the basis of which these dispositions are predicated. In support of this, it can be argued that one way of rebutting my statement would be to show that I was mistaken about the facts. The match would not have lit; it had already been used; or it was not dry. There is, however, the difficulty that I may very well subscribe to the conditional without knowing all the relevant facts. I may have forgotten or never have learned that oxygen has to be present. I may not know what it is about the constitution of matches that makes them combustible. And it is not plausible to say that I am implicitly making assertions which I am not competent to make explicit. I might, however, be held to be making the more general assertion that conditions in this instance are such that under conditions of that kind matches always light when they are struck. Alternatively, one might take the view that I was not asserting this but presupposing it. The advantage of this view would be that it would bring this type of example more into line with the other. Thus, in my previous example, one would not wish to maintain that when I said that the penny would come up heads I was implicitly asserting that it was biased, and there seems no good reason why one should change one's position on this point when the backing for the conditional is a universal generalisation rather than a statement of tendency.

If we treat this conditional also as the expression of a willingness to bet on the match's lighting, on the supposition that it is struck, it is natural to regard the fact that I am relying on a universal generalisation as strengthening my bet, so long as my presuppositions are true. But even on the assumption that my presuppositions are true, I am not betting on a certainty. My conditional statement does not follow from them. The premiss that under such and such conditions all matches which are struck light entails nothing whatever about any match which is not struck.

Suppose then that someone were to agree with me on all the facts but yet maintain that if this particular match had been struck it would not have lit. He gives no reason. He just insists that this case would have been an exception to the general rule. We can even imagine his saying this in every case in which the antecedent is not satisfied. Would he be saying anything false? Well, he would not be saying anything which contradicted any fact. He would not be representing anything as being other than it actually is.

Nevertheless, whether or not we chose to call it false, his statement would be unacceptable. And the reason why it would be unacceptable is that it conflicts with our method of arranging facts. It does not fit in with our conception of the way things work. The same rules govern our projection of past experience to imaginary as to future instances. In neither case do we think ourselves entitled to credit them with a different outcome unless we can give some reason why it should be different. And in this example I am assuming that no reason is given.

Let me put it in another way. Our picture of the world is a picture not only of actual events, or events which are believed to be actual, but also of imaginary events which branch out from the actual ones. Logically speaking, the requirement that our beliefs be factually true leaves us free to characterise the imaginary events in any way we please, but, in practice, we limit our freedom by adopting principles of construction which apply to the whole picture. An example would be the principle that things which have the same structure behave in the same way, whether the behaviour be actual or imaginary. It is as if we feared that to allow deviations on the branch line would lead to trouble on the main line. The imaginary events are regarded as being equipped to take the place of actual events and therefore as being subject to the same arrangement. This is not, indeed, althogether true of stories which are avowedly fictional, in the sense that they do not take off, in the same way, from actual situations. But even they are expected to have some verisimilitude; and verisimilitude primarily consists in adhering to the principles which govern our arrangement of facts.

The way in which we limit our imagination can be illustrated by an

example which I take from C. S. Peirce. 'We may ask', he says, 'what
prevents us from saying that all hard bodies remain perfectly soft until
they are touched, when their hardness increases with the pressure until
they are scratched.'[1] His first answer is that 'there would be no falsity
in such modes of speech' since 'they represent no fact to be different
from what it is; only they involve arrangements of facts which would
be exceedingly maladroit'. But in a later essay he argues that the arrangements are worse than maladroit. His reason is that the condition of the
object in question, say a diamond, is not 'an isolated fact'.[2] It may have
been tested for other properties which are associated by law with hardness. And even if it has not been tested, other diamonds have. The
point here is that we do not allow ourselves to attribute to an unexamined object a different structure from that which we know to be
possessed by others of its kind. And if we attribute the same structure
to it, then we do not allow ourselves to conceive of it, even in imaginary situations, as behaving in a different way.

So long as our arrangements are governed by what we take to be
laws, or even by strong statistical generalisations, which we feel entitled
to project, the question whether or not a non-truth-functional conditional deserves to be accepted is easy to decide, at least in particular
instances, however difficult it may be to formulate general rules. This
is, however, by no means always the case. Very often the grounds on
which one advances a non-truth-functional conditional are such that its
acceptability remains open to debate. This is especially so when human
behaviour is in question. 'If you stood for the post, you would be
elected.' 'If I had had the courage to sacrifice my bishop, I should have
won the game.' 'If Hitler had not invaded Russia, he would have won
the war.'

It may be instructive to consider how in a case of this kind the debate
might proceed. Confronted with the last example, one might argue:
Even if Hitler had not invaded Russia, he would still have lost the war:
Why? Because sooner or later the Americans would have come in and
in the long run the Germans would not have been able to resist them.
But would the Americans have come in if the Japanese had not attacked
them, and would the Japanese have gone to war if the Russians had not
been engaged with the Germans? Or again, it might be claimed that if
Germany had not been at war with Russia, then even if the Americans
had come in all the same, we should still not have won the war, since
the German divisions which were tied up in Russia would then have

---

[1] *The Collected Papers of C. S. Peirce*, v 403.   [2] Ibid., v 453.

been free to oppose our landing in the West. But then the atom bomb would have done for them. But if the Germans had not been so fully committed to a conventional war in Russia, might they not have developed it too? These may not all be equally good arguments, but they are all respectable and typical of the way in which such debates are conducted. Yet it is not at all easy to see how one is to decide between them. We move from one non-truth-functional conditional to another, almost without touching the springboard of a fact.

Nevertheless, there are facts in the background. Particular facts which are thought to justify dispositional statements about the character of certain politicians and the relative industrial capacities of the countries which they govern; and very weak generalisations such as the generalisation that armies with better equipment tend to beat armies with less good equipment, when their numbers are approximately equal. It may very well be that neither of the partners to such a dispute is able to assemble a set of facts which outweigh those which are adduced by his opponent, and in that case we have no call to say that either of them is right. On the other hand, one of the rival conditionals may find a backing which is relatively strong enough to make it acceptable. In such a case, what makes one set of facts outweigh another is that the generalisations which they exemplify are thought to be the more strongly projectible. And the point of asserting the conditional is just to direct attention to these facts and to the weight that should be attributed to them.

Where there is agreement on the facts, the usual reason for the maintenance of conflicting conditionals is that there is doubt or disagreement about the relative strength of different statements of tendency. It can, however, also happen that people maintain conflicting conditionals because they disagree about the implications of the divergence into fiction. A good example of this has been provided by Mr P. B. Downing.[1] He imagines a case in which two men, Jim and Jack, have quarrelled, and postulates first that if Jim were to ask Jack for help after they had quarrelled he would not get it, whereas otherwise he would, and secondly that while Jim would ask Jack for help if they had not quarrelled, he would not ask for it if they had. From these data, one party, Smith, straightforwardly draws the conclusion that if Jim were now to ask Jack for help, he would not get it, but his more subtle opponent, Jones, infers that he would get it, since it is assumed that he

[1] P. B. Downing, 'Subjunctive Conditionals: Time Order and Causation', *Proceedings of the Aristotelian Society*, LIX (1958–9).

would get it if there had been no quarrel and that it is only if there had been no quarrel that he would ask for it. Downing takes the view that one or other conclusion must be true, and himself sides with Smith, on the ground that the conditional on which Jones relies, 'If Jim were to ask Jack for help, they would not have previously quarrelled', is not a genuine subjunctive conditional but what he calls a subjunctive implication. 'Subjunctive implications,' he says, 'are roughly of the form "Circumstances are such that either not-p or q would be ful-filled".'[1] They are neutral with respect to the time-order of the events referred to in their antecedents and consequents, whereas subjunctive conditionals, according to Downing, are forward-looking in time. Thus, in his view, it is because causal statements entail subjunctive conditionals that we have to accept the rule that causes must precede their effects.

But even if it be granted that Jones is relying on a subjunctive implication, rather than a subjunctive conditional, why should this make his conclusion wrong? If I understand it correctly, Downing's answer to this is that when we raise a question about what would happen, in a given set of circumstances, if such and such an event E were to occur, we do not normally concern ourselves with the previous question whether in those circumstances the occurrence of E is possible. We are interested only in deciding what would ensue if E did occur, whether it could occur or not. So Jones's argument is at fault because it violates this rule. But while this may be a correct account of our ordinary practice, it does not strike me as a sufficient ground for rejecting arguments of Jones's type. And in this instance I should not wish to reject it. But this does not mean that I want to reject Smith's argument. If Downing assumes that one or other of them has to be rejected, it is because he takes a naïvely realistic view of non-truth-functional conditionals. Either Jim would help Jack, or he would not. But, as I see it, what we have here are two different excursions into fiction, both of which are acceptable. Smith takes the circumstances as they are and makes the supposition that Jim asks Jack for help, ignoring the fact that in the circumstances the supposition is not justified; he then gives the correct answer that Jim would not get help from Jack. Jones supposes the circumstances to be altered so that the further supposition that Jim asks Jack for help is justified, and on this basis he gives the correct answer that Jim would get his help. If in general we look more favourably on arguments of Smith's type, it is because we are more

[1] Ibid., p. 130.

concerned with limiting the area of fiction than in justifying our excursions into it. Thus I confidently say that if I were to play in a Grand Masters' chess tournament, I should lose every game. It could be argued that I should not be allowed to play in such a tournament unless I were myself a Grand Master; and in that case I probably should not lose every game. But the supposition that I am a Grand Master involves too great a departure from the facts to be of any interest.

It may be objected that what is significant here is not the comparative fidelity to fact but the question of time-order. Many suppositions, like the supposition that Hitler did not invade Russia, involve, if they are pursued, a very considerable departure from subsequent facts, but this does not deter us from making them. The facts which we are reluctant to tamper with are those that are temporally prior to the point of departure. But the answer to this is that when a supposition threatens to take us a great distance away from what we know to be the subsequent facts, we do not pursue it very far. We may argue about what would have happened if Hitler had not invaded Russia, to the point of imagining an entirely different outcome to the war, but we cannot seriously carry the argument much beyond that point, because we have already strayed so far from the facts that hardly any basis for an acceptable conditional remains. We feel a little more free to speculate about the future, because here we are not at variance with ascertained fact, but our ignorance of what is to come, and the limited character of the projections that we can reasonably make from our present basis of fact, impose similar restrictions upon the lengths to which our suppositions can usefully be carried.

It is this, I think, that explains why the non-truth-functional conditionals that enter into an argument are mostly forward-looking. There is nothing amiss with backward-looking conditionals as such. If it was reasonable to say before the last election that the Conservatives would come into power if they promised to reduce prices but not otherwise, then it is just as reasonable to say now that they would not have promised to reduce prices if they had not subsequently come into power. Admittedly, the second conditional sounds strange: we should ordinarily express it by saying that the Conservatives would not have come into power unless they had promised to reduce prices; but, whichever formulation is chosen, the same point is being made. What is in question here is a pairing of facts in a causal arrangement and it makes no difference in which order they are taken. The same applies, indeed, at least in the majority of cases, to our arrangement of imaginary events or events of which the factual status is uncertain; but because the speculations which interest us most are concerned with events of which the factual status is uncertain, and because these lie mainly in the actual or possible future, we tend to express all our arrangements both of actual and imaginary facts in a way that is temporally forward-looking. As I have tried to show elsewhere,[1] there is no logical asymmetry between the past and the future: there is some asymmetry, as is shown for example in the laws of thermodynamics, in the arrangement of facts on which causal judgements depend, but not enough to account for our making it a rule that the effect should not precede the cause. The explanation is to be found, I think, rather in the fact that we have so little knowledge of the future, compared with the extent of our knowledge of the past.

There remains the question what is needed for a non-truth-functional conditional to be acceptable; and my answer to this is a variant of that which has already been given by Nelson Goodman. I think that a conditional of this sort is acceptable if and only if it is supported more strongly than any of its rivals by a set of facts which include a true generalisation that we are willing to project. It might be thought that the requirement that the generalisation in question should be true was too strong, since, for all we know, a universal generalisation for which we have good evidence may have exceptions to it which we have not been in a position to discover; and yet we may still want to say that the conditionals which it sustains are to be accepted, considering the information that we have. But the answer to this is that we are allowing conditionals to be sustained not only by universal generalisations, but also by strong statements of tendency; and the function which is performed by a universal generalisation which is actually false may still be performed when the generalisation is reduced to a true statement of tendency.[2] The conditional which it sustains will thereby be weakened but not usually to the point where it ceases to be acceptable.

E. THE CONCEPT OF LAW

The requirement that these generalisations should be such as we are willing to project is also the key to the distinction between generalisa-

[1] See *The Problem of Knowledge*, pp. 192–8.
[2] See above, pp. 86–7.

tions of law and generalisations of fact. As I have argued elsewhere,[1] this distinction does not lie in the character of the generalisations themselves, but in our attitude towards them. So far as their factual content goes, there is no difference between them. A true generalisation of fact holds good in every instance in which its antecedent is exemplified and a true generalisation of law can do no more. The difference is that when one looks upon a generalisation as a generalisation of law one is willing to extend it to unknown and to imaginary instances in a way that one is not willing to extend any generalisation that one is treating only as a generalisation of fact.

This point has to be stated carefully, since it can be objected that if one believes a generalisation of fact to be true one thinks of it as holding in every actual instance; and when it comes to imaginary instances we must avoid relapsing into the circle of saying that a generalisation of law is one that supports an acceptable non-truth-functional conditional and that a non-truth-functional conditional is acceptable when it is supported by a true generalisation of law. Perhaps the best way of meeting these difficulties is to show how the distinction operates in a situation where the enumeration of the instances which fall under one generalisation is not known to be complete. Let the generalisation H be of the simple form 'for all x, if fx, then gx', and let there be an object O of which I believe truly or falsely that it has the property f. Then if I believe H to be true I shall indeed infer that O has the property g, whether I am treating H as a generalisation of law or as a generalisation of fact. All the same, if I am treating H only as a generalisation of fact the information that O has certain other properties will inhibit my inference in a way it will not be inhibited if I am treating H as a generalisation of law. For instance, if I believe that all those who are coming to the meeting will vote for the motion, my belief may not withstand the further information that one of those coming, whom I know to be corrupt, has been bribed to vote against it. On the other hand, no such information will weaken my belief in the proposition, which I believe to be the consequence of a generalisation of law, that all those who come to the meeting will be warm-blooded.

The point is not that we are more confident of the generalisations that we treat as generalisations of law than of those that we treat as generalisations of fact. On the contrary if we believe them at all, we are nearly always more confident of those that we treat as generalisations

---

[1] See my essay 'What is a Law of Nature?', *Revue Internationale de Philosophie*, xxxvi 2 (1936), reprinted in *The Concept of a Person*.

of fact, since we usually do not believe them until we have checked, or have good reason to believe that someone else has checked, what we take to be all the instances. It is that when we believe a generalisation which we treat as one of law, our belief in the invariable conjunction of its antecedent with its consequent is resistant to the ascription of other properties to an object which is presumed to satisfy the antecedent, whether the object's possession of these other properties be actual or imaginary. It is not quite correct to say that a belief in a generalisation of law is resistant to any enlargement of its antecedent; the additional properties must, for example, be logically compatible with the consequent, and there are complications which result from the fact that generalisations of law may interlock.[1] They are, however, no more than complications; even when they are allowed for, the distinction in our attitude remains, if only as a distinction of degree. As a distinction of degree, it caters for propositions which, as it were, fall between generalisations of law and generalisations of fact. For example, I do not regard it as a law of nature that all the men in the front row of my lecture audience should be wearing shoes, but local customs being what they are, I do not regard it as entirely accidental, in the way I should regard it as entirely accidental if it turned out that all the men in the front row were only sons. My belief in this last generalisation would not extend to a hypothetical acceptance of one that differed from it only in the circumstance that an extra man was added to the row, whereas my belief in the other would. My acceptance of an extension of the other would, however, not be resistant to such a supposition as that the audience contained members whose religious convictions or social nonconformism obliged them to go barefoot. My belief that they are all warm-blooded would evidently be resistant to any suppositions of this kind. It would only fail to withstand the hypothetical ascription to the persons concerned of physical properties, which were incompatible with those on which warm-bloodedness is thought to depend. In general, it can be said that the wider the range of properties to which our belief in a generalisation is, in this sense, resistant, the deeper the place that it occupies in our theoretical system.

At this point it may be objected that we have not after all escaped from circularity. For we have given an account of generalisations of law, and equally of non-truth-functional conditionals, in terms of our dispositions to adopt certain attitudes; and how are these dispositions themselves to be analysed except in terms of non-truth-functional

---

[1] See 'What is a Law of Nature', op. cit.

conditionals? The best answer that I can find to this objection is that we can explain what is meant by treating a generalisation as one of law, and so explain how a non-truth-functional conditional can be acceptable, by describing actual instances in which somebody makes the required projection. Having learned what it is to extend a generalisation to an undetermined or imaginary instance, we can comprehend the case where the generalisation which is extended is itself a generalisation about the process of extension. We can not, indeed, define the concept of natural law without making use of conditionals – for one thing, we want to allow for there being laws of nature which have not yet been discovered, and this requires us to conceive of there being true generalisation which we should treat as generalisations of law if we came to believe them; but if the idea of our being disposed to make projections can be regarded as being sufficiently explained by purely factual descriptions of actual human behaviour, this need not involve us in any vicious form of circularity.

F. CAUSALITY

In conclusion, I want to make a few brief remarks about the concept of cause. Like other statements which trade on the concept of law, causal statements are only partly factual. For the rest, they exhibit our arrangements of facts and the extensions of these arrangements to imaginary instances. Their factual content depends upon the level at which they operate. At the observable level, the factual content of a universal causal statement is that of the corresponding factual generalisation: the factual content of a singular causal statement consists in an assertion of the existence, in such and such a spatio-temporal relation, of the states of affairs which is conjoins, together with whatever generalisation is the basis for the conditional that one of these states of affairs would not in the circumstances have occurred without the other; this may be a universal generalisation, but it may also be a statement of tendency.

Not all universal generalisations of law are ordinarily classified as causal. For instance, functional laws, like the law connecting the temperature, volume and pressure of a gas, and taxonomic laws, like the laws that describe which insects are deciduous, would not be included. It may be disputed even whether the term is correctly used to cover such laws as the Newtonian laws of motion. This is, indeed, one of those matters with regard to which ordinary usage is not standardised and where an attempt to make it out in detail would not repay the labour. I shall, therefore, be content to distinguish four main types of case in which we commonly speak of causality.

The first class consists of the cases where a generalisation, whether of fact or of law, is explained by means of a wider theory, and some term which is intended to encapsulate the theory is taken to designate the cause of the phenomena with which the generalisation deals. It is in this large sense that we speak of the effects of gravitation or of electricity.

The second class consists of the cases in which the behaviour of an object is explained in terms of its composition or its structure, which is then said to be the cause of its behaving as it does. It is in this sense that we use the word 'because' when we make such a statement as that a piece of elastic stretches because it is made of rubber.

Thirdly, there are the cases in which a disposition, or a state of mind, is said to be the cause of its manifestations. Thus, a man may be said to yawn because he is bored, or to enter politics because he is ambitious. Such statements are not so trivial as they sometimes are thought to be. They fit the manifestation into a wider pattern.

Finally, there are the cases in which a causal statement links two states of affairs at the same observational or theoretical level. These are the examples most favoured in philosophical literature and their pre-eminence has led to the practice of representing causality as a relation between events. We shall, however, see that this is too narrow a con-ception even for this sort of case.

It is not always obvious to which of these classes a causal statement should be assigned. Thus, in a case where a glass is broken, the statement that it broke because it was brittle fits most naturally into my third category, but it could also be construed as referring to the composition of the glass. Or again, since the classes are not mutually exclusive, the statement that arsenic is poisonous can be taken both as referring to the composition of arsenic and as implying the existence of a regular con-nection between two states of affairs, namely that of someone's taking arsenic under such and such conditions and his subsequently suffering in such and such a characteristic way.

In the case of all statements of my fourth class, there is an implicit reference to a set of conditions, both positive and negative, which are also necessary for the production of the effect. This means that the state of affairs which is described as the cause is not thought to be sufficient for the effect to ensue, but only to be one of a set of factors which are taken to be jointly sufficient. The question then arises why one of these factors should be singled out from its fellows and alone dignified by the title of cause.

The answer to this is that, from a logical point of view, this procedure is, indeed, arbitrary, but it is not entirely capricious. There is some method to it. For example, it is usually possible to distinguish between what Professor Price has called standing and differential conditions[1] – that is to say, conditions which are relatively stable and conditions which come in as changes – and it is then the differential conditions that are singled out as causes. Thus, in the case of a forest fire, it is the spark that ignites the fire, and the wind that fans it, that are said to be the causes of the conflagration, rather than the state of the climate or the composition of the wood. Again, a differential condition is more likely to be singled out in this way if it consists in, or is believed to result from, some human action or omission. The flowers died because the gardener forgot to water them. This example also illustrates our tendency to fasten on to a condition which deviates from the norm. Presumably, the flowers would have survived if anyone had watered them, not just the gardener; but because the gardener was expected to do it, his omission is singled out. There is also the point that we feel entitled to blame the gardener as the one who had the responsibility of caring for the flowers and when human conduct is in question the choice of a cause is often linked with the purpose of allotting praise or blame.

Another merit of this example is that it shows the insufficiency of representing causality, even under my fourth heading, as a relation between events; for in no ordinary sense of the term could the gardener's inaction be classed as an event. But then we may be asked what are the terms of the causal relation if they are not events. The best rejoinder would be to refuse the question as implying a misunderstanding of the nature of causality. However, we could perhaps say that the terms of the causal relation were facts, so long as we made it clear that all we meant by this was that every causal statement could be represented as offering an explanation of the truth of one proposition in terms of the truth of another.

With regard to the question how the states of affairs to which these propositions refer are spatio-temporally related, the only established rule appears to be that the effect should not precede the cause; and even for this, as we have seen, there seems to be no good reason beyond the

---

[1] See H. H. Price, *Perception* (1932), p. 70.

fact that we have less knowledge of the future than we can claim to have of the past. It is also fairly generally assumed that in the cases where the cause can be represented as an event which precedes the effect, the two events must be temporally contiguous, or very nearly so. We are, indeed, willing to allow that events which are distant from one another in time are causally related, but we think of them as being joined by a causal chain and then single out the last link in the chain as the effective cause. Thus, we believe that a present memory is causally dependent upon the past experience of which it is a memory, but we also tend to think of the past experience as leaving some physical trace within the brain and of the memory as effectively arising from the response of this relic to an immediately preceding, usually unidentified stimulus. In cases where we are short of materials to bridge the gap, we content ourselves with a fictitious bridge, as when we connect present memories with the experiences of childhood by postulating that traces of these experiences remain in the unconscious. When spatial relations are in question, we are less uneasy about the idea of action at a distance, but even so we are apt to bring in fictitious entities, like forces, to bridge the gap. The explanation is, I think, that we cherish a model of causality which is derived from primitive experiences of manipulating objects and we tend therefore to be dissatisfied with any account of a causal process that does not exhibit a mechanism which includes at least a substitute for physical contact. Once again, it is hard to see any logical justification for this requirement. In no case are we doing anything more than systematically correlating different states of affairs; and there seems to be no good reason why a theory in which the states of affairs which are correlated are spatio-temporally adjacent should be regarded as being *pro tanto* superior to one in which they are saptio-temporally remote.

In the case where two concurrent states of affairs are symmetrically correlated, in the sense that each is taken to be necessary and sufficient for the other, one might expect that they would be regarded as mutually determinant. Nevertheless, this is not always so. For example, those who accept the hypothesis that some contemporary state of a man's brain is necessary and sufficient for the existence of any of his mental states conclude from this that the mind is causally dependent upon the brain: they do not conclude that the brain is dependent upon the mind, although it follows logically, if the existence of such and such a brain state is necessary and sufficient for that of some mental state, that the converse also holds. The reason for this discrepancy is, I think, that we can give an account of the workings of the brain which fits into a wider physical system, and mental occurrences are given only a dependent status because their correlation with the brain introduces them into this wider system in which they play no essential part. Admittedly, mental states are themselves thought capable of having physical effects, but the hypothesis that they are perfectly correlated with states of the brain gives us the assurance that there is always a physical agent available to do the same work. On the other hand, there are many physical facts employed in the wider system for which no mental replacements are available. It is true that we also correlate mental states with one another, but these correlations are too fragmentary and too limited in their range to furnish a rival to the wider physical system.

The reference to a wider context also accounts for the cases in which we treat one correlate as a sign of its partner rather than its cause. A falling barometer is thought to presage rain and not to cause it, because we possess a more comprehensive theory which accounts both for the fall in the barometer and for the subsequent rain as dependent facts. Similarly, though the height of a telegraph-pole and the length of the shadow which it casts are mutually determinant, when considered by themselves, the fact that we can account for the height of the pole without taking into consideration the length of the shadow, whereas we can not account for the length of the shadow without taking into consideration the height of the pole, leads us to think of the shadow as dependent on the pole and not the other way around. There is also the point that we can do things to the pole which will make a difference to the shadow, whereas we can not do anything directly to the shadow which will make a difference to the pole.

I said earlier that every causal statement, to be acceptable, needed the backing of some true generalisation, but that it was not necessary that these generalisations should be anything more than statements of tendency; they did not need to be, or even be taken for, causal laws. The best illustration of this is to be found in the singular causal statements that we make about human conduct. If I say, for example, that my child is crying because he has not been allowed to stay up to watch a late television programme, I do not make this assertion on the basis of any belief in such universal generalisations as that children invariably cry when they are not allowed to stay up to watch television or even that my own child invariably does so. I know these generalisations to be false. It may well be that there is some true generalisation to the effect that under such and such conditions children of such and such a character always cry, which would be applicable to the present case: indeed, if the conditions are specified so narrowly that they obtain only in the present case this must be trivially true. But even if there is a projectible generalisation of this kind which would justify us in saying, if we knew of it, that the child's behaviour was causally determined, I am not relying on it when I make my causal statement; I am not even claiming that it exists. The position is rather that I believe there to be a limited number of ways in which children come to cry, one of which is their being disappointed of some expected pleasure. This belief may be expressed in a set of tendency statements: if children are so disappointed they quite often cry; if children suffer physical pain they quite often cry; if children are harshly spoken to they quite often cry; and so forth. The generalisations may take a more specific form as concerning only children of such and such a type or perhaps only one particular child, and the conditions also may be more precisely specified. Then if I find, in the given situation, that the antecedent of only one of this set of tendency statements is satisfied, I pick out the state of affairs which exemplifies it as the cause of that which is exemplified by their common consequent. The relatively weak tendency statement on which I am relying acquires the force of a causal explanation through the defection of all its admissible competitors.

   This example will also serve to show that it is not strictly correct to equate a singular causal statement 'q because p' with the unfulfilled conditional 'if not p then not q'. The reason for this is that in asserting 'q because p' I do not embark upon any story about what would be happening if p were false. For instance, in saying that the child is crying because he has not been allowed to watch television I do not exclude the possibility that if he were watching television, the programme itself would move him to tears. Nevertheless, the equation of 'q because p' with 'if not p then not q' is correct to the extent that taking the situation as it is, and not looking beyond it to the possibilities admitted by supposing not-p, which might in their turn favour q, I do pair q with p in a way that does license the unfulfilled conditional under this restriction. The point could be made by saying that if he had been allowed to go on watching television he would not be crying unless something else had occurred to make him cry. This statement is not trivial, since it associates his crying with only one of the facts that actually obtained. This is an instance of a weak association, holding its place through lack of competition, but the same considerations apply to cases where the singular causal statement is backed by a universal generalisation, since for the most part, at least in everyday discourse, the set of jointly sufficient conditions, in which the one referred to as the cause is a necessary element, is not conceived to be uniquely sufficient for the production of the effect. It is only when a plurality of sufficient conditions is excluded that we obtain the perfect equation of a causal statement with the corresponding unfulfilled conditional.

There are those who would say that the child's disappointment was the reason for his crying rather than its cause. But once it is admitted that causal statements may be supported by generalisations of tendency and not only by universal generalisations, the ground for contrasting motives with causes disappears. There might be some point in maintaining the distinction if the reliance on generalisations of tendency was confined to the sphere of human, or human and animal, behaviour, but this is not so. For example, our assignment of causes for change in the weather is based on nothing stronger than generalisations of tendency. No doubt we believe that there are causal laws to be discovered in this field, but the point here again is that we make causal statements without having discovered these laws and without referring to them.

The familiar objection that motives and intentions can not be causes because they are not distinct events from the acts with which they are correlated is simply met by remarking that even if they are not distinct events, the facts of their occurrence are distinct. The proof of this is that when it is said that a man intends to do such and such an act, or that he has such and such a motive, or such and such a reason for doing it, the question whether he actually does it is logically left open. But if, as in cases of this kind, we are able to associate distinct facts through a true, and on the occasion dominant, generalisation of tendency, then, as has been shown, we have an adequate basis for an attribution of causality.

To explain an action in terms of the motive from which it was done does not preclude the belief that it can also be explained in another way – for instance, in terms of a physiological theory. There is no logical reason why one and the same fact should not occur in different explanatory arrangements. I think, however, that if the physiological explanation were available, we should attach more importance to it, regard it as telling the deeper story, first because it would presumably rest on stronger generalisations, and secondly, once again, because it would fit into a wider pattern.

The upshot of this discussion is that in a certain sense causes are what we choose them to be. We do not decide what facts habitually go together but we do decide what combinations are to be imaginatively projected. The despised savages who beat gongs at solar eclipses to summon back the sun are not making any factual error. It is a true generalisation that whenever they beat the gongs the sun does shine again, and if they always keep up the ceremony, it is also a true generalisation that the sun comes out again only when they beat the gongs. If we despise them, it is because they tell a fictitious story about what would happen if they did not beat the gongs, which we do not accept. They see what goes on as well as we do; it is just that we have a different and, we think, a better idea of the way the world works.

Index